Beyond Solar Cells: The Expanding World of Perovskite Optoelectronics

Michael

First Printing, 2024

TABLE OF CONTENTS

Chapter *1*

Background, Objective and Structure of the book

Optoelectronic devices either convert light to electricity (photodetectors and solar cells) or convert electricity to light (light-emitting diode). These devices, together with light modulators and optical memory, have ushered an era of internet of things (IoT).[1] The IoT has an increasing market size since 2012 and is expected to have robust growth, as shown in **Figure 1.1**.

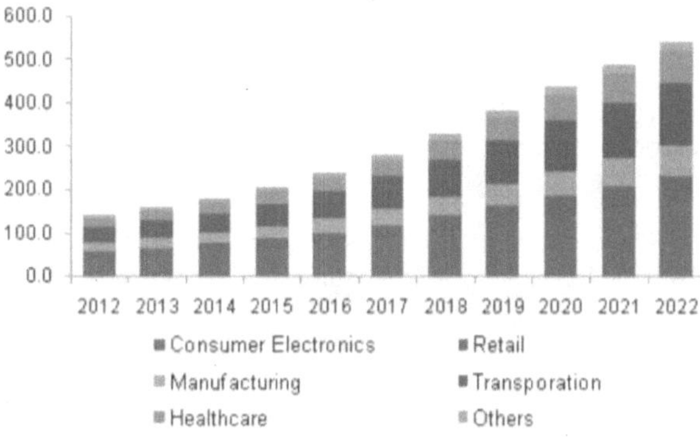

Figure 1.1 North America IoT market revenue by applications. 2012- 2022 (USD Billion).[1]

Optoelectronics, combining the optical information conversion with the capability of integration with conventional microelectronic devices, plays a fundamental role in the new era of IoT, ranging from sensing, processing, energy supply, and computing as shown in **Figure 1.2**.[2]

Figure 1.2 Optoelectronics plays a fundamental role in the new era of IoT, ranging from sensing, processing, energy supply, and computing.[3]

After a decade of investigation in hybrid organic-inorganic lead halide perovskite materials and their related optoelectronics, the first generation of perovskite-based solar panels has been introduced into the commercial market.[4] The commercial success of the perovskite solar cells is on the basis of the in-depth investigation of the perovskite materials, including their inherent properties and fabrication technologies. At present stage, conventional perovskites such as $MAPbI_3$ and $MAPbBr_3$ are proven to be feasible and capable to fabricate other optoelectronics due to their outstanding photo-electric conversion efficiency, which allows the rapid development of these devices, including light-emitting diode (LED),[5] photodetector,[6] phototransistor,[7] and optical memory.[8]

However, the performance of optoelectronic devices based on the conventional perovskite ($CH_3NH_3PbI_3$) is limited by ion migration,[8] the mobility of the carriers, and the light absorption in the near-infrared (NIR) region, *etc*.[9] The perovskite is an orthorhombic crystal at room temperature, and the iodine atoms appear to be mobile in the crystal, which leads to a series of problems, including ion migration, carrier scattering.[10] Thus, the difficulty in achieving high-performance perovskite-based optoelectronic devices is to suppress the ion migration. From another perspective, using other functional materials to promote charge separation and conduction is an alternative strategy.

The main objective of this work is to explore and optimize the integration of hybrid organic-inorganic lead halide perovskite with other functional materials. The integrations of different dimensional materials were fabricated using various methods to optimize the performance of the optoelectronic device. Then, the corresponding mechanism was explored.

This book is organized in the following structure:

Chapter 2 provides the fundamental background of perovskite and related literature review. Basic knowledge of optoelectronic devices used in this book is demonstrated, including solar cells, photodetector, thin-film transistor (TFT), and resistance switching memory.

Chapter 3 provides detailed methods for material growth and characterization.

Chapter 4 demonstrates the fabrication of perovskite solar cells embedded with zero-dimensional Au-dimer. We exploited the application of nanostructures of Au nanorod-nanoparticle dimers with structural darkness to enhance the light-harvesting and the performance of perovskite solar cells.

Chapter 5 demonstrates a photoconductor made of perovskite films and WS_2 monolayers. We used WS_2 monolayer to separate electron and holes and to assist carrier transport, which results in an enhancement of photodetectivity and response speed.

Chapter 6 demonstrates a phototransistor made of bulk heterojunction of made of perovskite films and SWCNT. We utilize sorted semiconducting single-walled carbon nanotubes to enhance the performance of TFTs, enabling mixed-dimensional solution-processed electronics with high mobility and low voltage operation.

Chapter 7 reports the development of hybrid memtransistor, modulable by multiple physical inputs (optical and electrical inputs), using hybrid perovskite and conjugated polymer heterojunction channels. The devices exhibit non-volatile but highly reversible conductance modulation

Chapter 8 finalizes the conclusions of this book, and discuss prospects for the perovskite optoelectronics.

Reference

1 *IoT Market Analysis By Component (Devices, Connectivity, IT Services, Platforms), By Application (Consumer Electronics, Retail, Manufacturing, Transportation, Healthcare) And Segment Forecasts To 2022*, <https://www.grandviewresearch.com/industry-analysis/iot-market> (2015).

2 Kwon, J. *et al.* Recent progress in silver nanowire based flexible/wearable optoelectronics. *J Mater Chem C* **6**, 7445-7461 (2018).

3 Gipson, M. *Sensors – The Lifeblood of the Internet of Things*, <https://semielectronics.com/sensors-lifeblood-internet-things> (2017).

4 Mathews, I., Kantareddy, S. N., Buonassisi, T. & Peters, I. M. Technology and Market Perspective for Indoor Photovoltaic Cells. *Joule* **3**, 1415-1426 (2019).

5 Lin, K. B. *et al.* Perovskite light-emitting diodes with external quantum efficiency exceeding 20 per cent. *Nature* **562**, 245-+ (2018).

6 Miao, J. L. & Zhang, F. J. Recent progress on highly sensitive perovskite photodetectors. *J Mater Chem C* **7**, 1741-1791 (2019).

7 Li, F. *et al.* Ambipolar solution-processed hybrid perovskite phototransistors. *Nat. Commun.* **6**, 8238 (2015).

8 Guan, X. W. *et al.* Light-Responsive Ion-Redistribution-Induced Resistive Switching in Hybrid Perovskite Schottky Junctions. *Adv. Funct. Mater.* **28**, 1704665 (2018).

9 Yue, L. Y., Yan, B., Attridge, M. & Wang, Z. B. Light absorption in perovskite solar cell: Fundamentals and plasmonic enhancement of infrared band absorption. *Sol Energy* **124**, 143-152 (2016).

10 Senanayak, S. P. *et al.* Understanding charge transport in lead iodide perovskite thin-film field-effect transistors. *Sci Adv* **3**, e1601935 (2017).

Chapter 2

Literature Review and Fundamental Knowledge in Perovskite Semiconductor and Optoelectronics

This chapter provides fundamental background knowledge in perovskite semiconductor and relevant optoelectronic devices that are used in this book. The intrinsic physical and chemical properties that relate to the remarkably high performance are demonstrated. After the material fundamentals, the basic knowledge of optoelectronics is provided, including device configuration, operation mechanisms, and the figures-of-merit to evaluate the device performance. Some literature reporting the development history of the perovskite semiconductors are reviewed, followed by the corresponding optoelectronic devices with state-of-the-art performance.

2.1 Introduction

Halide perovskites have multiple unique advantages as electronic materials. Metal halide perovskites have been attracting enormous attention for applications in optoelectronic devices. Examples of these optoelectronic devices include photodetectors,[1] light-emitting devices,[2] and field-effect transistors.[3] In 2015, methylammonium lead halide attracted intensive attention as prototypical dye-sensitized solar cells with certified power conversion efficiencies exceeding 25%.[4] The remarkable semiconducting properties of halide perovskites that enable and trigger a wide range of applications have been intensively investigated in recent years. And the potential of halide perovskites in optoelectronics is very promising. Strong inter-band transitions give rise to high light absorption,[5] which is accompanied by bimolecular recombination below the Langevin limit.[6] There is a photon-recycling property in perovskite thin films, allowing for efficient light absorption.[7] In addition to these beneficial properties, perovskites possess very competitive performance amongst solution-processed electronic materials.

2.2 The Crystal and Electronic Band Structure of Perovskites

A perovskite structure is any material with the same type of crystal structure as calcium titanite (CaTiO3), typically with space group Pm3m. The unit cell of lead halide perovskites is demonstrated in **Figure 2.1a**. The cubic structure of the perovskite unit cell consists of five atoms (A represents ammonium ions, B represents a metal cation, and C represents halide anions). Currently, the most-studied lead halide perovskites usually have MA+ anions for the A site and lead as the B site.

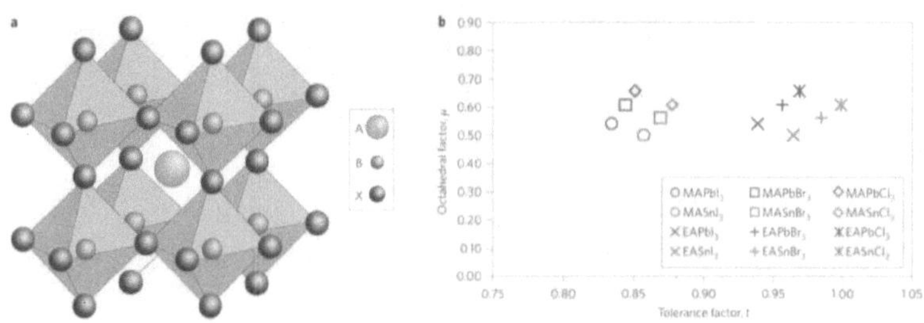

Figure 2.1 a. Cubic perovskite crystal structure. **b.** Calculated t and μ factors for 12 halide perovskites.[8]

Despite the various choices of A, M, and X ions from organic to inorganic, there is one empirical equation governing the allowed radii of the ions: $R_A + R_X = t\sqrt{2}(R_M + R_X)$, where R_A, R_X, and R_M are the ionic radii of corresponding ions in AMX3, and t is the tolerance factor. t should satisfy: $0.8 < t < 1$, to form a stable three-dimensional (3D) perovskite structure.[8] Some commonly used ions to construct perovskite structure are shown in **Figure 2.1b**. As an example, for the most widely-studied organolead halide perovskites, in which M = Pb (R_M = 1.19 Å) and X = Cl, Br or I (R_X = 1.84, 1.96, 2.20 Å, respectively), we can calculate that R_M should not exceed ~2.6 Å. Although hundreds of perovskites have been studied, the organolead halide perovskites show the most attractive performance in optoelectronics. Therefore, this book will focus on organolead halide perovskites.

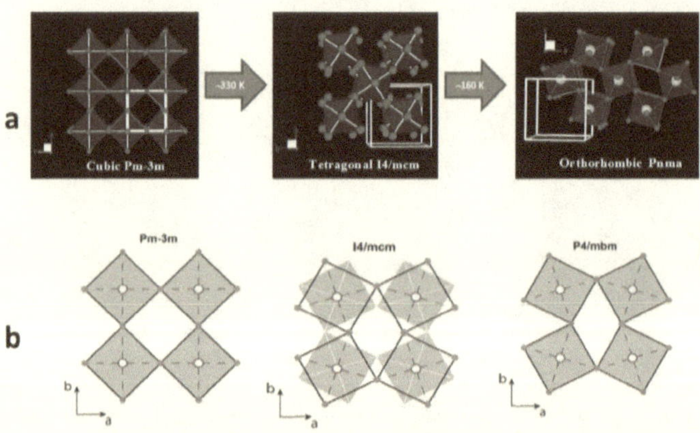

Figure 2.2 The crystal structures adopted by $MAPbI_3$. (a) The PbI_6 octahedra are blue, and the iodine atoms are red. The cubic to tetragonal transition is due to the R4+ rotational distortion mode, and the tetragonal to orthorhombic transition is primarily associated with a combination of the R4+ and M3+ distortion modes. (**b**) The relative rotations of neighboring layers of PbI_6 octahedra along the c axis are shown as filled green squares and unfilled black squares.[9]

As **Figure 2.2a** shows, The $MAPbI_3$ perovskite has a reversible orthorhombic to tetragonal phase transition at around 160 K. And a phase transition from cubic to tetragonal is at about 330 K.[9] The corresponding space groups and the PbI_6 octahedra are demonstrated in **Figure 2.2b**. Many reports have also revealed the phase transition from spectroscopy methods, such as Wu, *et al.* studied the revolution of two excitonic peaks confirms the structural transition from the orthorhombic phase to the tetragonal phase at the critical temperature of 160 K.[10]

The remarkable physical properties are the key factors to enable and trigger the wide range of applications investigated in recent years. On the one hand, as shown in **Figure 2.3a** that strong inter-band transitions give rise to high absorption, which accompanies the bimolecular recombination below the Langevin limit. On the other hand, long carrier relaxing time (**Figure 2.3 b,c**), high carrier mobility (**Figure 2.3 d**) and low trapping density (**Figure 2.3e**),[11] together with photon-recycling property in perovskite thin films,[7] lead to an extraordinarily long carrier diffusion length,[12] which paves the way for the widespread use of perovskite optoelectronics.

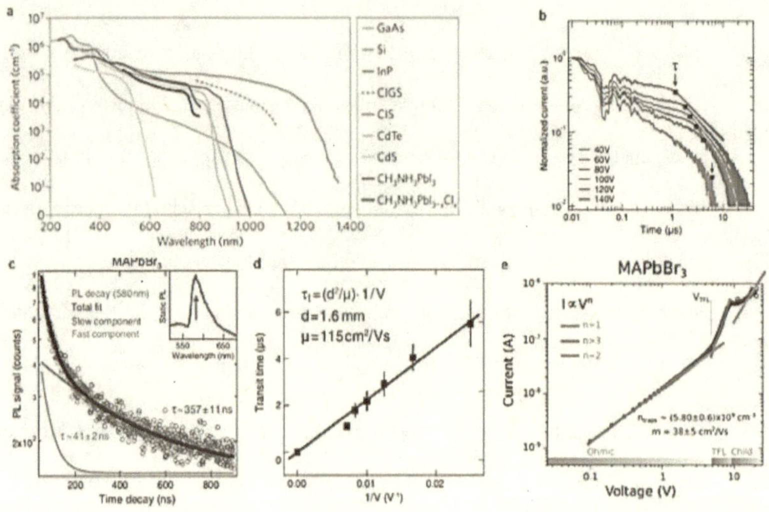

Figure 2.3 The physical properties by MAPbI₃. (a) Absorption coefficient of $CH_3NH_3PbI_3$.[8] (b) Time-of-flight traces showing the transient current after photoexcitation at time t = 0 in a bilogarithmic plot.[11] (c) PL time decay trace on a MAPbBr3 crystal at l = 580 nm.[11] (d) Linear fit of transit time versus inverse voltage.[11] (e) Characteristic I-V trace (purple markers) showing three different regimes for $MAPbI_3$.[11]

As shown in **Figure 2.4,** the 1.5 eV bandgap of $MAPbI_3$ is formed between the unoccupied Pb ***p*** orbital and the occupied I ***p*** orbital. These advanced theoretical calculations help explain some fundamental mechanisms for the superior properties of lead halide perovskites compared to other popular photovoltaic (PV) materials. As shown in

Figure 2.4 a,b, Yan and his co-workers have demonstrated two main important properties of $MAPbI_3$ perovskite based on theoretical calculations. (1) The much higher optical absorption of $MAPbI_3$ than GaAs is due to the density of states (DOS) derived from the Pb *p* orbital in the lower conduction band (CB) of the halide perovskites is significantly higher than that of GaAs owing to a more dispersed ***s*** orbital from Ga. (2) The strong ***s–p*** anti-bonding coupling in $MAPbI_3$ perovskite leads to small effective masses for both electrons and holes, which is helpful for efficient solar cells with a p–i–n configuration. Based on advanced DOS calculation, as shown in **Figure 2.4 c**, the effective mass of exciton is ~0.104 m_e. A light effective mass gives rise to strong tunneling and less scattering compared to conventional semiconductors, such as Si has the effective mass of exciton of 0.4 m_e, and Ge has 0.08 m_e.

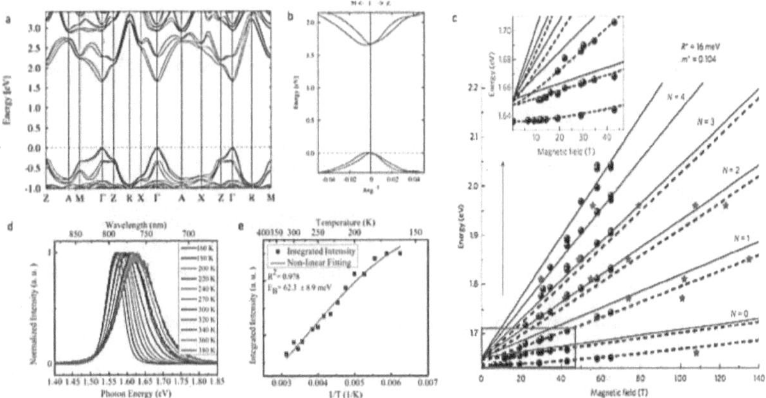

Figure 2.4 a. Theoretical calculation of the band structure of the $MAPbI_3$ perovskite.[13] **b.** The direct bandgap with 1.65 eV of the $MAPbI_3$ perovskite.[13] **c.** Effective exciton mass of the perovskite measured by magneto-optical analysis.[14] **d.** temperature-dependent PL

measurement from 160 K to 380 K. **e.** fitting for extracting the binding energy of the $MAPbI_3$ perovskite.[10]

Besides the suitable direct bandgap and the effective mass, the ultra-long carrier diffusion length and low trap density are the other factors that contribute to the high mobility of perovskites. As shown in **Figure 2.4 d,e**, an exciton binding energy of 62.3 ± 8.9 meV and optical phonon energy of 25.3 ± 5.2 meV were estimated by temperature-dependent photoluminescence spectroscopy.[10]

2.3 Fundamentals of Optoelectronic Devices

2.3.1 Junctions

In general, there are many different types of junctions that can be applied to multiple applications. Specifically, p-n junctions have a wide application in the design of electronic and optoelectronic devices. For example, the p-n junctions consist of n- and p-type semiconductors from either the same material (homojunction) or different materials (heterojunction). The silicon p-n junction has been widely applied to the industry, including diodes that perform functions such as rectification and switching. In addition to the p-n junctions, metal-semiconductor junction, and metal-insulator-semiconductor (MIS) junctions have a wide potential to be applied in recent electronics. For example, one of the most important types of the metal-semiconductor junction is Schottky junction, which is used to rectify currents. The MIS junctions are widely used in MOSFET (metal-oxide-semiconductor field-effect transistor), which we will discuss in the next section (**2.3.4**). In the following section, we will discuss the physics of the junction in the equilibrium and the carrier mechanisms in the non-equilibrium.

2.3.2 P-N junctions

The simplest semiconductor junction diode is formed by the combination of a pair of p-type and n-type semiconductors. Many optoelectronic devices are constructed based on the junction. Therefore, knowledge of the p-n junction is crucial for the operation of the devices.

When any two subsystems are in thermal and diffusive contact, there is no net current flow in equilibrium. Therefore, we see that the gradient of Fermi levels (E_F) of the subsystems must be zero in equilibrium. However, the energy level for two different p-type and n-type materials are intrinsically different. If we place the p- and n-type material into contact, a new equilibrium state is established, and the Fermi level of the whole system must line up flat across the junction, as shown in **Figure 2.5**. To maintain the flat Fermi level, a depletion region (x_p to x_n in the figure) will form due to the diffusion and drift of the carriers. Most optoelectronic devices operate based on the carrier flows when the junctions are illuminated, and non-equilibrium carriers (photo-generated carriers) are injected.

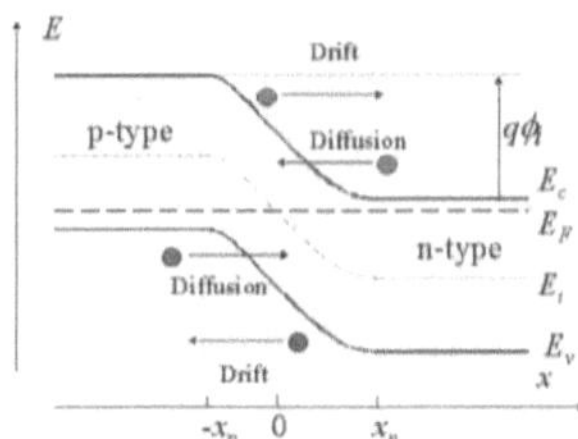

Figure 2.5 Energy band diagram of a p-n junction in equilibrium. A flat Fermi level (E_F) is formed due to the diffusion and drift of the carriers.[15]

One important application of the PN junction is the rectification of currents. The rectifying operation of PN junction is originated from the formation of the depletion region, in which the diffusion of carriers from the p-type (or n-type) semiconductor recombine with the opposite carriers in the n-type (p-type) layer, resulting in the formation of the charge depletion region, and a built-in barrier is spontaneously established. The schematic of the space charge region and the built-in electric field is shown in **Figure 2.6**. When an external bias is applied to the PN junction, the built-in barrier will rectify the current. A reverse bias (*i.e.*, high voltages on n-type and low on p-type) increases the height of the built-in barrier (ΔV), thus the current across the junction suppressed. On the contrary, a forward bias (*i.e.*, low voltages on n-type and high on p-type) decreases the height of the (ΔV) and affiliates the carrier flow.

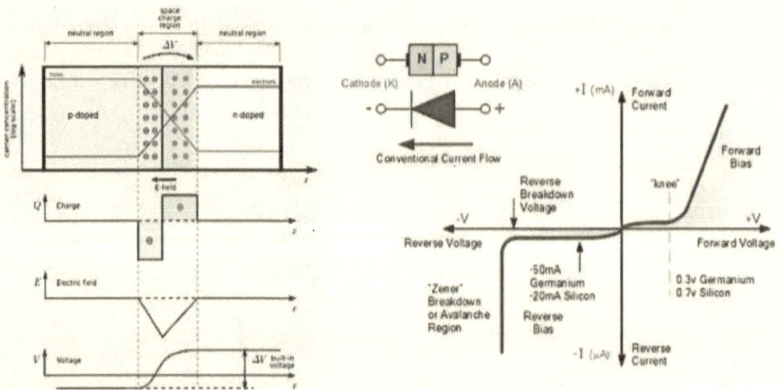

Figure 2.6 A p–n junction in thermal equilibrium with zero-bias voltage applied and current rectification with bias.[16,17]

In terms of perovskite PN junction, the PVK is an intrinsic ambipolar semiconductor. However, the ion migration in the perovskite dopes itself after applying a high voltage polling, as shown in **Figure 2.7.** This intriguing characteristic leads to the development of many switchable devices, such as giant switchable photovoltaics,[18] resistive switching memory,[19] and memtransistors.

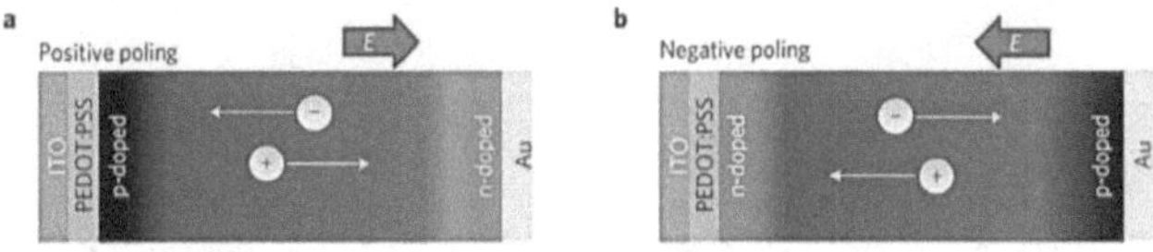

Figure 2.7 self-doping in the perovskite films under a voltage polling.[18]

2.3.3 Solar Cells

Photovoltaics is an important optoelectronic technology, used for the harvesting of solar energy and the conversion of said energy to electricity. This technology has the potential to supplement global energy consumption and overcome the global environmental problem caused by the burning of fossil fuels. The development of the PN junction in 1949 inspired the invention of the solar cells in 1954. Since then, solar cells have been developed through four generations with crystalline silicon, polysilicon, other crystalline materials (CdTe, GaAs, and CIGS), and dye-sensitized solar cells.[20] Recently, perovskite solar cells have attracted enormous attention due to their power conversion efficiency (PCE) up to 25.2%.[4] A development of the best PCE with various generations of solar cells is shown in **Figure**

2.8. The application of solar cells is now an important and integral part of IoTs, which usually require small sizes and wide distributions. The integration of a solar cell is used to provide energy to the IoT devices.[21]

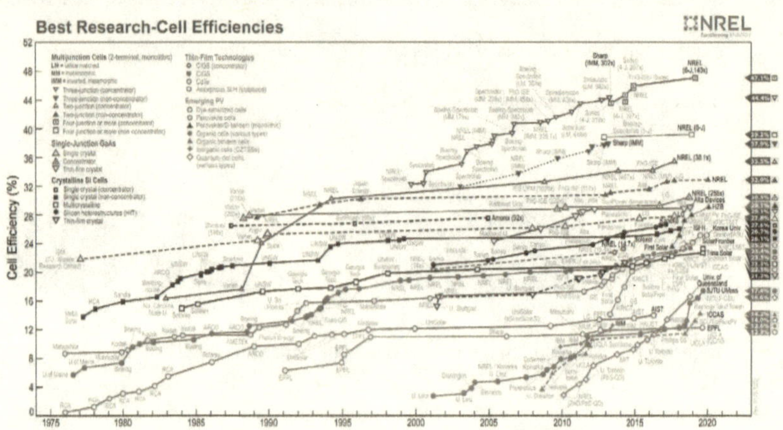

Figure 2.8 Best research photovoltaics efficiency on various generation technologies.[22]

Typically, a solar cell consists of a PN junction. Incident sunlight is absorbed in an absorber layer and generates electron-hole pairs. These photo-generated carriers are separated within the PN junction and collected by the hole transport layer (HTL) and the electron transport layer (ETL), respectively. A schematic working principle of solar cells is given in **Figure 2.9.**

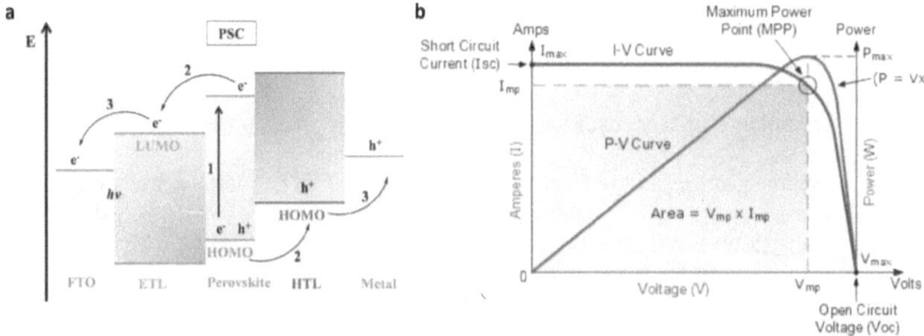

Figure 2.9 a. Band diagram and working principle of perovskite solar cells.[23] b. current-voltage characteristics of a solar cell under illumination.[24]

In order to assess the quality of a solar cell, some critical parameters are of importance. As shown in **Figure 2.9 b**, the current-voltage characteristic of a solar cell under illumination is widely used. Power conversion efficiency (PCE) is defined as the ratio of output power to input power (PCE = P_{out}/P_{in}). More specifically, short circuit current density (J_{sc}) reflects the quantum efficiency of a solar cell, which can be used to evaluate the capability of the absorber layer. Open circuit voltage (V_{oc}) reflects the collection capability of carriers, which significantly depends on the bandgap of the PN junction. Loss and recombination of the carriers at interfaces can be seen from the loss of V_{oc}. In general, to maximize the output power, both J_{sc} and V_{oc} must be made as large as possible. Fill factor (FF) is used to define the power extraction efficiency and is expressed as[25]

$$FF = \frac{I_m V_m}{I_{sc} V_{oc}} \tag{2.1}$$

2.3.4 Thin-Film Transistors

Field-effect transistors (FETs) are the workhorse of modern microelectronics. Several important devices exploit the field effect in their operation, such as metal-oxide-semiconductor field-effect transistors (MOSFET), metal-semiconductor field-effect transistors (MESFET) and junction field-effect transistors (JFET). In general, an integrated circuit (IC) chip is constructed from billions of MOSFETs, operating with different functions from memory to computing. In this section, the fundamentals of the FET working principle are presented, and the literature of perovskite-based FETs are reviewed.

Arguably, the most widely used semiconductor field-effect transistor is the MOSFET. The MOSFET is a three-terminal device. (*i.e.,* source, drain and gate terminals), in which the drain current is controlled by the application of gate voltages (field effect). The electric field is provided by the gate electrode with a metal-insulator-semiconductor (MIS) structure. A schematic device structure is shown in **Figure 2.10 (a)**

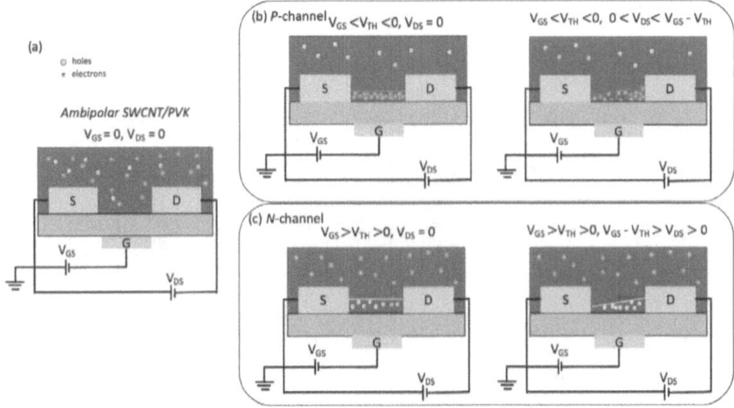

Figure 2.10 (a)schematic of ambipolar SWCNT/PVK TFT. (b)Working principle of an ideal ambipolar SWCNT/PVK heterostructure TFT.

In perovskite TFT, before a bias is applied to the device, the ambipolar SWCNT/PVK channel is unperturbed and intrinsic.[26] Charge carriers (both electrons and holes) distribute evenly throughout the thin film as shown in **Figure 2.10 (a).** As shown in **Figure 2.10 (b),** when a negative gate voltage ($V_{GS} < V_{TH} < 0$) is applied, an electric field is created and causes the formation of an accumulation layer of holes at the semiconductor-dielectric interface. Subsequent application of increasing bias ($0 < V_{DS} < V_{GS} - V_{TH}$) to the drain electrode causes the charges to travel laterally along the channel between the source and the drain electrodes. As of n-type channel in **Figure 2.10 (c),** a positive gate voltage ($V_{GS} > V_{TH} > 0$) is applied, the electric field induced the accumulation of electrons at the interface. When a positive source-drain voltage ($V_{GS} - V_{TH} > V_{DS} > 0$) applied, electrons transport laterally along the channel.

Depending on the spatial location of the gate electrode, source electrode, and drain electrode, the TFT can be further categorized into four subclasses[27] (*i.e.*, TGTC, TGBC, BGTC, and BGBC), namely, top-gate (-TG) and bottom-gate (-BG) with the combination of top-contact (-TC) and bottom-contact (-BC). The schematics of these subclasses are shown in **Figure 2.11.** Furthermore, a recent technology called FinFET (Fin field-effect transistor) has been invented, which built on a substrate where the gate is placed on multiple sides of the channel or wrapped around the channel, forming a 3D structure.[28] The utilization of the FinFET breaks the bottleneck and reduces the size of nodes down to 7 nm. In general, all of the device configurations have their advantages and disadvantages. For example, the BG TFT is facile and allows light to directly interact with the semiconductor layer. However, the TG TFT sustains a better semiconductor quality due to the absence of electrodes when depositing a semiconductor layer. In some cases, post-deposition annealing is preferred to attain a desirable crystal structure.

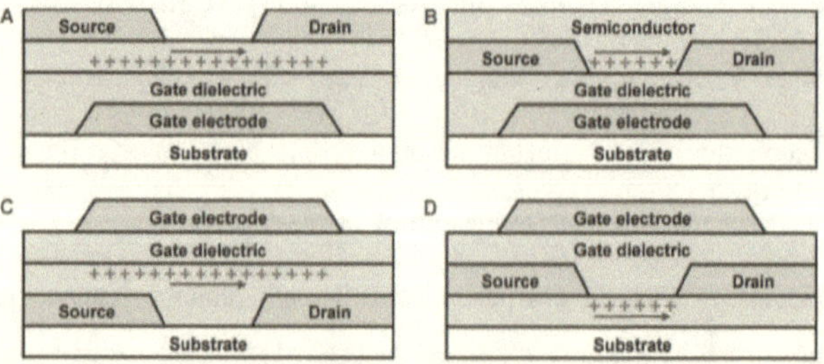

Figure 2.11 schematic cross-section of the four typical thin-film transistor structures. (a) bottom-gate top-contact staggered TFT (BGTC). (b) bottom-gate top-contact coplanar TFT (BGTC). (c) top-gate bottom-contact staggered TFT (TGBC). (d) top-gate top-contact coplanar TFT.[27]

The working principle of the field-effect transistor is modulating the source-drain current by application of a gate voltage. This modulation is realized by controlling the injection of carriers near the semiconductor/dielectric interface. To access the performance of a transistor, the current-voltage characteristics of the source vs. drain electrodes and gate vs. drain electrodes are of importance (*i.e.*, output characteristic and transfer characteristic). The typical static characteristics of the TFTs are shown in **Figure 2.12**, where both linear and saturation regions are shown.

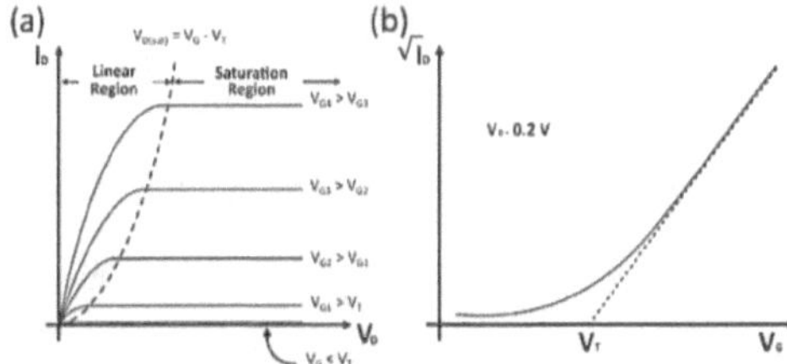

Figure 2.12 (a) output and **(b)** transfer characteristic of a generic n-type transistor.[29]

The output curve of an n-type TFT, as shown in **Figure 2.12**(a), can be separate into two regions as divided by a dashed line (V_G-V_T) in the figure.

Linear region. When $V_G > V_T$ and $V_{DS} < V_G$-V_T. The transistor is turned on, and a conductive channel has been created, which allows the current to flow between source and drain electrodes. The TFT operates like a resistor with ohmic contact, and the conductivity of the channel is controlled by the gate voltages. Thus, in this region, the current modulation can be described as[30]

$$I_D = \mu_n C_{ox} \frac{W}{L}\left((V_G - V_T)V_{DS} - \frac{V_{DS}^2}{2}\right) \tag{2.2}$$

where μ_n is the charge-carrier effective mobility, W is the channel width, L is the channel length, and C_{ox} is the gate oxide capacitance per unit area.

Saturation region. When $V_G > V_T$ and $V_{DS} > V_G - V_T$, the channel has been created, which allows current between the drain and source. Since the drain voltage is higher than the source voltage, the electrons spread out, and the channel pinch-off. Thus, current flows in a broader region than at the semiconductor/dielectric interface. Due to the lack of the channel region near the drain electrode, the drain current is now weakly dependent upon drain voltages and controlled primarily by the gate voltage. The current modulation can be described as[30]

$$I_{DS} = \mu_{sat} C_{ox} \frac{W}{2L}(V_G - V_T)^2 \tag{2.3}$$

where μ_{sat} is saturation mobility.

The carrier mobility is the most important parameter to evaluate the performance of a transistor. In addition to the carrier mobility, the transfer characteristics permit quantitative assessment of a large number of important parameters, including ON/OFF ratio (I_{ON}/I_{OFF}), threshold voltage (V_T), turn-on voltage (V_{on}), subthreshold swing (SS), interface trap density (D_{it}). These parameters can be extracted from the transfer characteristics as follows:[31,32]

$$SS = \left(\max \frac{\partial \log I_{DS}}{\partial V_G}\right)^{-1} \tag{2.4}$$

$$D_{it} = \frac{1}{q}\left(\frac{qSS}{2.3kT} - 1\right) C_{ox} = \frac{C_{ox}}{q} \Delta V_T \tag{2.5}$$

Specifically, in the transistors based on perovskites, PVKs are also expected to accelerate the development of solution-processed electronics. **Table 2.1** provides a list of representative studies on PVK-based thin-film transistors (TFTs). In one of the first studies, Mitzi *et al.* reported a $(PEA)_2SnI_4$ TFT with mobility of 0.6 cm^2/Vs.[33] More recent work has focused on $MAPbI_{3-x}Cl_x$ TFTs with reported mobilities in the range of 1.01-1.24 cm^2/Vs.[3] Several efforts have attempted to improve TFT performance by optimizing PVK composition,[34] fabricating hybrid bilayers,[35] and using single-crystal microplates.[36] A particularly powerful approach is to interface PVK with high-mobility nanomaterials such as graphene and carbon nanotubes (CNTs). However, these devices have thus far suffered from low ON/OFF ratios and high threshold voltages, limiting their electronic applications.

Table 2.1 TFT performance of the state-of-the-art researches

Materials	Dielectric	Mobility (cm^2/Vs)	Threshold voltage (V)	ON/OFF ratio	inverter Gain	Reference
$(PEA)_2SnI_4$	SiO_2	0.29 – 1.7	~ -5	~ 10^6		33
$MAPbI_{3-x}Cl_x$	SiO_2	1.0 – 1.3	5 – 10	~ 10^5		3
$MAPbI_3$/PEIE	Cytop/ SiO_2	3 – 5	~ -3	~ 10^5		34
$MAPbI_3$ nanoplate	SiO_2	1 – 2.5	~ 15	Nearly 10^6		36
$MAPbI_3/WSe_2$	SiO_2		~ 20	10^6		35
$Cs_x(MA_{0.17}FA_{0.83})_{1-x}Pb(Br_{0.17}I_{0.83})_3$	SiO_2	2.02 – 2.39	~ 50	10^4 – 10^6	23 (V_{DD}=80 V)	37
$MAPbI_{3-x}Cl_x$/unsorted CNTs	SiO_2	108.7–595.3	~ 2	10^2		26

2.3.5 Photodetectors

A photodetector is an optoelectronic device that absorbs optical signals and converts it to electrical signals, which usually manifests as a photocurrent. The photodetectors are the most important application of compound semiconductor materials in optoelectronic devices, and are widely used in optical communication systems that can be used by a telephone, a computer, or other terminals at the receiving end. In these applications, the detector must be designed to ensure the low-loss signal conversion, the speed of response, and the low noise. Therefore, we will discuss the category, the mechanisms of photodetectors in this section, and review recent researches on perovskite-based photodetectors.

Many direct-bandgap semiconductors are considered to be the desirable materials for making photodetectors as their low energy loss, such as GaAs, InP, GaInAs, GaN, PbS, *etc.* Depending on the energy of bandgaps, different semiconductors can be used to detect photons from the infrared, visible to ultraviolet, and X-ray of the spectrum. Perovskites have a wide range of efficient light absorption from the near-infrared to gamma-ray. Therefore, they have been used as an active layer to fabricate photodetectors.

Depending on the configuration of the devices, photodetectors can be classified into four main types, including photoconductors, photodiodes, phototransistors and, avalanche photodiodes, as shown in **Figure 2.13.**[1] Photodetectors are also classified into intrinsic and extrinsic types. An intrinsic photodetector usually detects the light of wavelength close to the bandgap of the active semiconductor. Upon illumination, the photoexcited electron-hole pairs are separate within the semiconductor. Diffusion and drift effects then change the conductivity of the materials, arising from the changes in carrier mobility or density. And an extrinsic photodetector detects light of energy smaller than the bandgap energy. Absorption of photos excites an electron from the deep energy in the bandgap to the conduction band, which is very useful in the detection of the far-infrared.

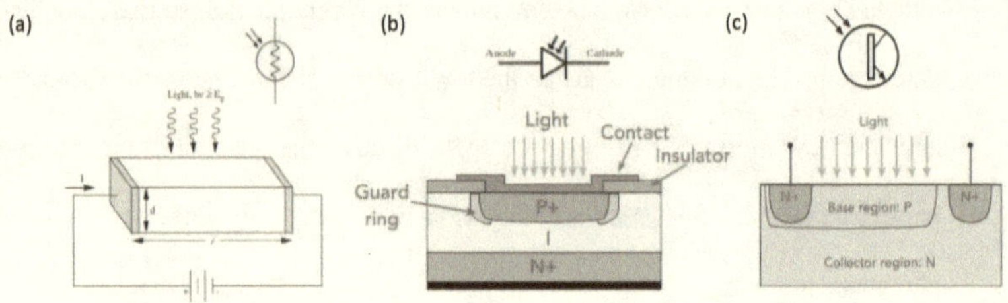

Figure 2.13 schematic device structure and electronic symbols of three different photodetectors. (a) photoconductor, (b) photodiodes, and (c) phototransistors.[38]

Figure of merits[39]

Quantity	Symbol/acronym	Unit	Definition
Responsivity	R	$A\,W^{-1}$	Photocurrent flowing in a detector divided by incident optical power
Dark current Dark-current density	I_d J_d	A $A\,cm^{-2}$	Current (density) flowing in the absence of illumination. In an unbiased photodiode at steady-state, this is equal to zero.
Quantum efficiency External quantum efficiency	QE EQE	%	In a photodiode, the ratio of photocurrent (in electrons per second) to photon fluence incident on the device (photons per second). Related to responsivity via the photon energy: $R = QE/E_{photon}$(where E_{photon} is measured in electronvolts). In a photoconducto the term QE is sometimes used synonymously with absorbance.
Internal gain	G_{int}	Unitless	In a photoconductor, the ratio of photocurrent (in electrons per second) to photon absorbed.
External gain	G_{ext}	Unitless	In a photoconductor, the ratio of photocurrent (in electrons per second) to photon incident. $G_{ext} = QE \times G_{int}$
Noise current	I_{noise}	$A/\sqrt{Hz}$	The random root mean square fluctuation in current when bandwidth is limited to 1 Hz.
Noise-equivalent power	NEP	$W/\sqrt{Hz}$	The minimum detectable power, that is, the optical signal in watts at which the electric signal-to-noise ratio in the detector is equal to unity (0 dB), when bandwidth is limited 1 Hz.
Normalized detectivity	D^*	$cm\sqrt{Hz}/W$ (Jones)	A measure of detector sensitivity that enables comparison even when detector area A and bandwidth B are different. $D^* = (\sqrt{AB})/NEP$.

Reference

1 Miao, J. L. & Zhang, F. J. Recent progress on highly sensitive perovskite photodetectors. *J Mater Chem C* **7**, 1741-1791 (2019).

2 Lin, K. B. *et al.* Perovskite light-emitting diodes with external quantum efficiency exceeding 20 per cent. *Nature* **562**, 245-+ (2018).

3 Li, F. *et al.* Ambipolar solution-processed hybrid perovskite phototransistors. *Nat. Commun.* **6**, 8238 (2015).

4 McMeekin, D. P. *et al.* A mixed-cation lead mixed-halide perovskite absorber for tandem solar cells. *Science* **351**, 151 (2016).

5 Herz, L. M. Charge-Carrier Mobilities in Metal Halide Perovskites: Fundamental Mechanisms and Limits. *Acs Energy Lett* **2**, 1539-1548 (2017).

6 Davies, C. L. *et al.* Bimolecular recombination in methylammonium lead triiodide perovskite is an inverse absorption process. *Nat. Commun.* **9** (2018).

7 Pazos-Outon, L. M. *et al.* Photon recycling in lead iodide perovskite solar cells. *Science* **351**, 1430-1433 (2016).

8 Green, M. A., Ho-Baillie, A. & Snaith, H. J. The emergence of perovskite solar cells. *Nat. Photonics* **8**, 506-514 (2014).

9 Whitfield, P. S. *et al.* Structures, Phase Transitions and Tricritical Behavior of the Hybrid Perovskite Methyl Ammonium Lead Iodide. *Sci Rep-Uk* **6**, 35685 (2016).

10 Wu, K. W. *.e..* Temperature-dependent excitonic photoluminescence of hybrid organometal halide perovskite films. *PCCP* **16**, 22476-22481 (2014).

11 Shi, D. *et al.* Low trap-state density and long carrier diffusion in organolead trihalide perovskite single crystals. *Science* **347**, 519-522 (2015).

12 Stranks, S. D. *et al.* Electron-Hole Diffusion Lengths Exceeding 1 Micrometer in an Organometal Trihalide Perovskite Absorber. *Science* **342**, 341-344 (2013).

13 Amat, A. *et al.* Cation-Induced Band-Gap Tuning in Organohalide Perovskites: Interplay of Spin-Orbit Coupling and Octahedra Tilting. *Nano Lett.* **14**, 3608-3616 (2014).

14 Miyata, A. *et al.* Direct measurement of the exciton binding energy and effective masses for charge carriers in organic-inorganic tri-halide perovskites. *Nat. Phys.* **11**, 582-594 (2015).

15 Zeghbroeck, B. V. *Principles of Semiconductor Devices*, <https://ecee.colorado.edu/~bart/book/book/chapter4/ch4_2.htm> (2011).

16 *p-n junction*, <https://en.wikipedia.org/wiki/P–n_junction> (

17 *PN Junction Diode*, <https://www.electronics-tutorials.ws/diode/diode_3.html> (

18 Xiao, Z. G. *et al.* Giant switchable photovoltaic effect in organometal trihalide perovskite devices. *Nat. Mater.* **14**, 193-198 (2015).

19 Guan, X. W. *et al.* Light-Responsive Ion-Redistribution-Induced Resistive Switching in Hybrid Perovskite Schottky Junctions. *Adv. Funct. Mater.* **28**, 1704665 (2018).

20 *Solar cell*, <https://en.wikipedia.org/wiki/Solar_cell> (2020).

21 Kwon, J. *et al.* Recent progress in silver nanowire based flexible/wearable optoelectronics. *J Mater Chem C* **6**, 7445-7461 (2018).

22 NREL. *Best Research-Cell Efficiency Chart*, <https://www.nrel.gov/pv/cell-efficiency.html> (2020).

23 Marinova, N., Valero, S. & Delgado, J. L. Organic and perovskite solar cells: Working principles, materials and interfaces. *J. Colloid Interface Sci.* **488**, 373-389 (2017).

24 *Solar Cell I-V Characteristic*, <http://www.alternative-energy-tutorials.com/energy-articles/solar-cell-i-v-characteristic.html> (2020).

25 Qi, B. Y. & Wang, J. Z. Fill factor in organic solar cells. *PCCP* **15**, 8972-8982 (2013).

26 Li, F. *et al.* Ultrahigh Carrier Mobility Achieved in Photoresponsive Hybrid Perovskite Films via Coupling with Single-Walled Carbon Nanotubes. *Adv. Mater.* **29**, 1602432 (2017).

27 Klauk, H. Organic thin-film transistors. *Chem. Soc. Rev.* **39**, 2643-2666 (2010).

28 Jurczak, M., Collaert, N., Veloso, A., Hoffmann, T. & Biesemans, S. Review of FINFET technology. *2009 IEEE International SOI Conference*, 3-6 (2009).

29 *Difference Between Depletion MOSFET vs Enhancement MOSFET*, <https://www.rfwireless-world.com/Terminology/Depletion-MOSFET-vs-Enhancement-MOSFET.html> (2020).

30 Faber, H. *et al.* Heterojunction oxide thin-film transistors with unprecedented electron mobility grown from solution. *Sci Adv* **3**, e1602640 (2017).

31 Nourbakhsh, A., Zubair, A., Joglekar, S., Dresselhaus, M. & Palacios, T. Subthreshold swing improvement in MoS2 transistors by the negative-capacitance effect in a ferroelectric Al-doped-HfO2/HfO2 gate dielectric stack. *Nanoscale* **9**, 6122-6127 (2017).

32 Fang, N., Toyoda, S., Taniguchi, T., Watanabe, K. & Nagashio, K. Full Energy Spectra of Interface State Densities for n- and p-type MoS2 Field-Effect Transistors. *Adv. Funct. Mater.* **29**, 1904465 (2019).

33 Mitzi, D. B. Solution-processed inorganic semiconductors. *J. Mater. Chem.* **14**, 2355-2365 (2004).

34 Senanayak, S. P. *et al.* Understanding charge transport in lead iodide perovskite thin-film field-effect transistors. *Sci Adv* **3**, e1601935 (2017).

35 Cheng, H. C. *et al.* van der Waals Heterojunction Devices Based on Organohalide Perovskites and Two-Dimensional Materials. *Nano Lett.* **16**, 367-373 (2016).

36 Wang, G. M. *et al.* Wafer-scale growth of large arrays of perovskite microplate crystals for functional electronics and optoelectronics. *Sci Adv* **1**, e1500613 (2015).

37 Yusoff, A. B. *et al.* Ambipolar Triple Cation Perovskite Field Effect Transistors and Inverters. *Adv. Mater.* **29**, 1602940 (2017).

38 Notes, E. Photodiode Technology. (2020).

39 Konstantatos, G. & Sargent, E. H. Nanostructured materials for photon detection. *Nat. Nanotechnol.* **5**, 391-400 (2010).

Chapter 3

Experimental Techniques

This chapter introduces the experimental methods and material preparation used in this book. The whole procedure includes thin-film deposition and characterization, electrode deposition, and optoelectrical measurement.

There are many methods used to deposit perovskite thin-film, such as sol-gel spin coating, thermal evaporation, and pulse laser deposition. In this chapter, we will introduce the sol-gel spin coating with multiple steps, post-deposition engineering, and thermal evaporation, which have been extensively used in the experiments.

In order to ensure the quality of the thin films, various characterization techniques have been utilized. For example, X-ray diffraction, scanning electron microscope, atomic force microscope, photoluminescence spectroscope, and Raman spectroscope, *etc*. For the precise patterning of electrodes, photolithography is commonly used.

3.1 Thin Film Growth Methods

3.1.1 Sol-Gel Spin Coating

There are different recipes for perovskite deposition, among which spin coating is the most facile and commonly used method. We chose the one-step spin coating method and followed with an anti-solvent dripping during the deposition.

In chapter 4, we explored rotation evaporation to recrystallize the precursors and solvent engineering during the deposition. Methylammonium iodide (CH_3NH_3I) was synthesized following; Hydroiodic acid (0.227 mol, 57 wt% in water, Aldrich) was slowly dropped into methylamine (0.273 mol, 40% in methanol) with ice-bath stirring at 0 °C for 2 h. After dropping, rotation evaporation was carried out at 50 °C for 1 h. Then the precipitate was dissolved in methanol and recrystallized by diethyl ether for three times and dried in a vacuum oven for 8 h. In order to maintain the same thickness of perovskite films, a high concentration precursor was first synthesized. 3.6 M CH_3NH_3I and 1.2 M $PbCl_2$ (molar ratio 3:1) was dissolved in anhydrous N,N-dimethylformamide (DMF) to produce a high concentration mixed halide perovskite precursor.[1]

A compact TiO_2 layer was spin-coated (2000 rpm, 30 s) using 0.15 M titanium isopropoxide and baked at 200 °C for 2 mins, followed with a post-annealing at 550 °C for 30 mins. After the substrates cooled down, a mesoporous TiO_2 layer was deposited by spin coating (2000 rpm, 45 s) diluted TiO_2 paste. The layers were sintered at 550 °C for 30 mins. The perovskite precursor solution (with and without Au dimers) was then spin-coated on m-TiO_2 at 2000 rpm for 60 s and thermally treated for 1 h, which is conducted in a glove box. A 68 mM spiro-OMeTAD solution in chlorobenzene containing 55 mM tert-butylpyridine and 9 mM lithium bis(trifluoromethylsyfonyl)imide (LiTFSI) salt was spin-coated the perovskite films at 2000 rpm, 45 s.[1]

In chapter 6, we explored solvent engineering during the deposition of mixed-cation mixed-halide perovskite. The Si/HfO_2 substrates were treated with oxygen plasma before film deposition. Au (80 nm) source (S) and drain (D) electrodes were deposited by thermal evaporation through a shadow mask, defining a channel with a length of 50 μm and a width of 1000 μm. A PEIE solution (0.4 w.t.% in 2-methoxyethanol) was spin-coated on the electrodes at 5000 rpm for 30 s. The substrates were then heated at 100 °C for 10 minutes, and followed with ultrasonication in DI water for 10 minutes. A thin layer of PEIE (~5 nm) was deposited after drying the substrates in nitrogen and annealing at 100 °C for 5 minutes in a glovebox. The mixed solution was then spin-coated on the substrates at 2000 rpm for 10 s. A two-step spin coating was performed at 6000 rpm for 30 s, followed by 50 μL of anti-solvent (toluene) dripping during the last 10 s, after which the devices were annealed at 100 °C for 1 hour in a glove box to reduce the charge traps and improve the Ohmic contact between the active layer and electrodes.

3.1.2 Thermal Evaporation

Thermal evaporation is the most commonly used physical deposition methods. As shown in **Figure 3.1,** a target source material is heated in a crucible. When the material reaches its boiling temperature, vapor flux is generated. By controlling the input power with a feedback system (PID), the vapor flux can be set to a certain rate, which ensures the homogeneous deposition of the target material. However, due to the limit of the boiling temperature, most oxides are not able to evaporate with this method.

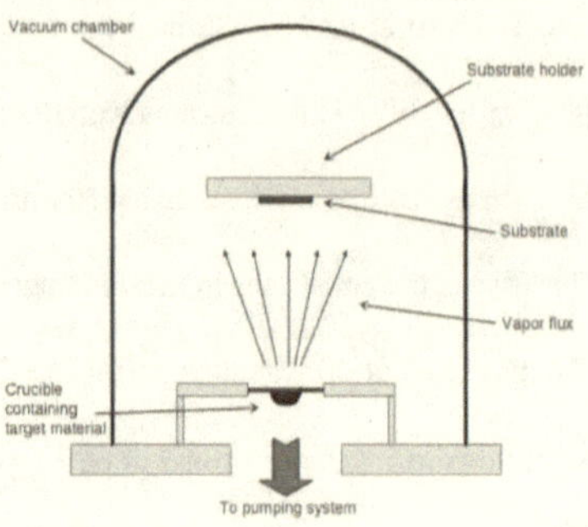

Figure 3.1 Schematic of the thermal evaporation.[2]

In chapter 5, we explored the thermal evaporation to deposit perovskite physically, in order to attain a van der Waals heterostructure with WS_2. The $CH_3NH_3PbI_3$ perovskite films were deposited onto the WS_2 monolayers using the sequential vapor deposition method. As-prepared WS_2 monolayer films were loaded into a thermal evaporation chamber (base pressure is below 1×10^{-4} Pa) to deposit PbI_2 films. The evaporation temperature is 152 °C, and the evaporation rate was 0.4 Å/s. After the PbI_2 film thickness reached 150 nm, samples were put on a hot plate in an atmosphere-controlled chamber, with CH_3NH_3I powder surrounding the substrate, and the PbI_2 films were converted into $CH_3NH_3PbI_3$ at a temperature of 120 °C. After two hours of phase conversion, one hour of post-annealing was carried out with DMF vapor to passivate the perovskite surface.[3]

3.1.3 Atomic Layer Deposition

In order to tackle the limitation of the thermal evaporation, oxides are deposited with other methods, such as the magnetron sputter and the atomic layer deposition (ALD). We used the ALD to deposit the dielectric layer (HfO) in transistors. A schematic is given in **Figure 3.2.**

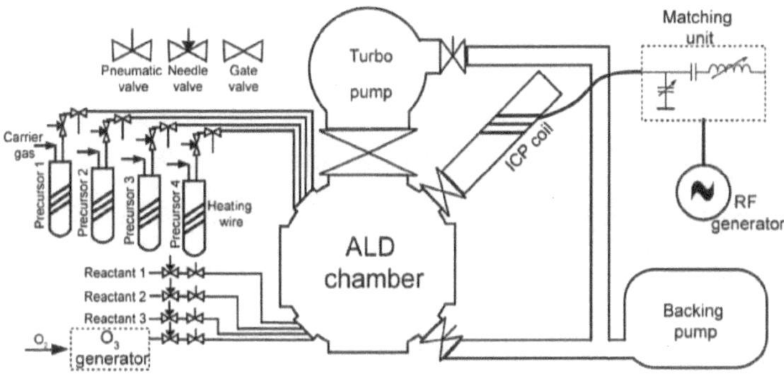

Figure 3.2 schematic mechanism and equipment of the ALD.[4]

3.2 Thin Film Characterization

3.2.1 X-ray Diffraction

Crystals are a regular periodic array of atoms, and X-rays can be considered as electromagnetic waves. When the incident X-ray diffraction elastically scattered by the atomic planes. The reflect angle follows Bragg's law:

$$2d \sin\theta = n\lambda \tag{3.1}$$

Here, d is the spacing between diffracting planes, θ is the incident angle, n is an integer, and λ is the wavelength of the X-ray. Thus, the X-ray diffraction is commonly used to identify the crystal and access their crystallinity.[3]

In addition, there is a relationship between the grains size and the full width at half maximum (FWHM) of the XRD patterns, therefore grains size can be roughly estimated by using the Scherrer equation:

$$\tau = \frac{K\lambda}{\beta \sin\theta} \tag{3.2}$$

where λ is the beam wavelength, K is a constant, β is the FWHM, and θ is the Bragg angle. τ is the mean size of the crystalline domains, which may be smaller or equal to the grain size. K varies with the shape of the domain, which is typically about 0.9.[5]

In this work, X-ray diffraction (XRD) of perovskite films and hybrid films were measured by a Bruker D8 ADVANCE diffractometer with Cu $K\alpha$ (λ = 1.5406 Å) radiation.

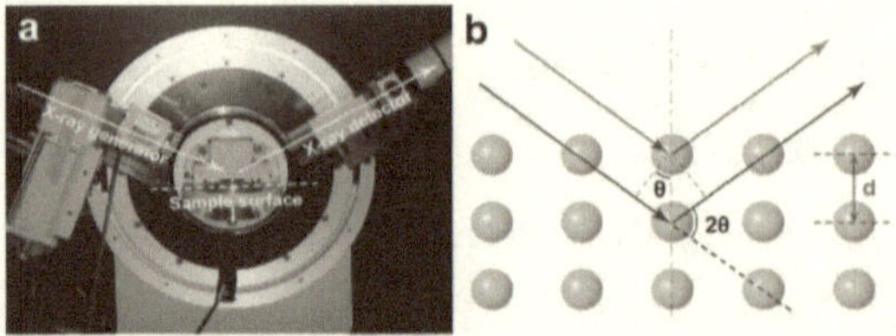

Figure 3.3(a) Photograph of a Bruker D8 Advanced X-ray powder diffractometer. (b) Schematic illustration showing X-rays interact with crystalline lattice.[6]

3.2.2 Scanning Electron Microscope

A scanning electron microscope (SEM) is a type of electron microscope that images a sample by scanning the surface with a focus electron beam, as shown in **Figure 3.4**. The electron interacts with the surface of the sample, providing information about the topography and composition of the sample. All SEM images in the book are measured with Nano Nova, and Quanta 600 scanning electron microscope provided by FEI in the core lab of KAUST.

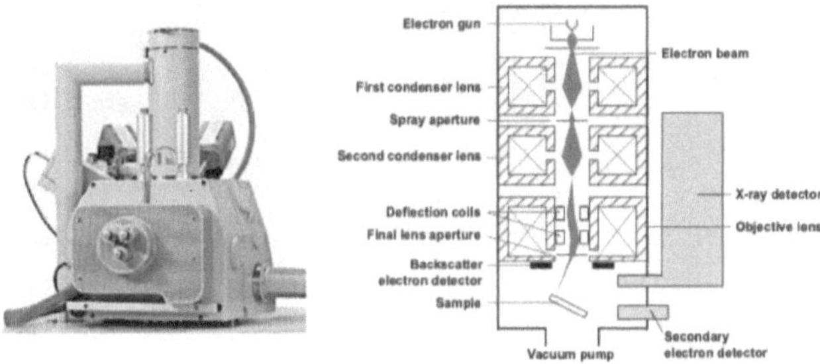

Figure 3.4 a photo of SEM setup provided by FEI and a schematic imaging mechanism of the SEM.[7]

3.2.3 Atomic Force microscope

Atomic force microscope (AFM) is a type of scanning probe microscopy, used to image the topography and surface potential of the sample surface, with the Z-resolution on the order of atom size. As shown in **Figure 3.5**, the AFM consists of a cantilever with a sharp probe at the end, which interacts with the sample surface. A position-sensitive photodetector signals the displacement of the laser beam, which provides information about the interaction force between the cantilever and sample surface.

In this book, Atomic force microscope (AFM) measurements were carried out with a Dimension Icon (Bruker) in air and in the dark. KPFM measurement was carried out with a scanning probe microscope (Solver next, NT-MDT), and OSCM-PT probes (Veeco) were used in the KPFM measurements.

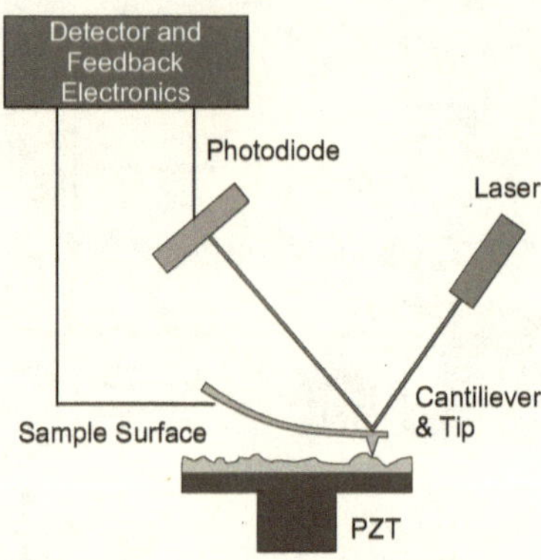

Figure 3.5 A schematic of the AFM.[8]

3.2.3 Raman Spectroscope

Raman spectroscope is a spectroscopic technique used to determine the vibration mode of molecular. The vibrational energy states of each molecular are different from each other, as shown in **Figure 3.6**. Thus, the Raman spectroscope is widely used to identify molecules.

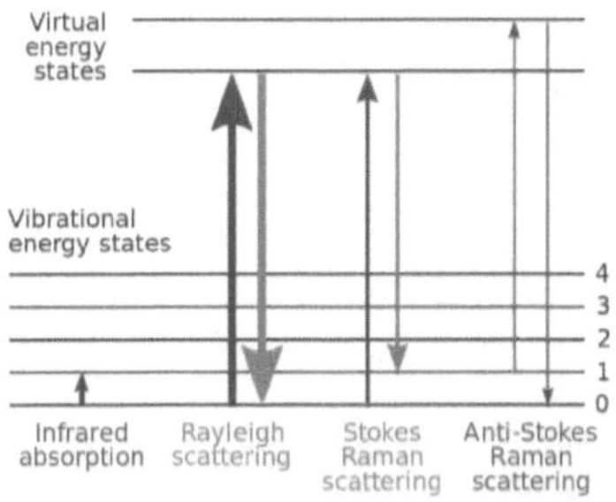

Figure 3.6 vibrational energy states involved in Raman spectra.[9]

In this book, Raman spectroscope was used to determine the layer number of WS_2. Monolayer WS_2 has a different fingerprint Raman spectrum with that of multiple layers. The Raman spectra of pristine WS_2, WS_2/perovskite hybrid sample, and perovskite films on sapphire substrates were obtained by confocal Raman microscopy systems (Witec Alpha 300), with the excitation wavelength of 532nm. The grating of PL spectrometer was 600g/mm, and the grating of Raman spectrometer was 1800g/mm.[3]

3.2.4 UV-Vis Spectroscope

Ultraviolet-visible spectroscopy (UV-Vis) is a kind of light absorption spectroscopy used to investigate the absorption efficiency of materials. In a certain region of the electromagnetic spectrum, atoms and molecules undergo electronic transitions. Specifically, by analyzing the absorption spectrum of semiconductors, the absorption edge determines the bandgap of the semiconductors.

In this book, UV-Vis spectroscopy is not only used to determine the bandgap of perovskites, but also to demonstrate the enhancement of light harvest by the addition of Au dimers. The absorption spectra of pristine WS_2 film, perovskite film, and hybrid systems were obtained by an Agilent Cary 5000 UV-Visible-NIR spectrometer.[3]

3.2.5 Photoluminescence Spectroscope

Photoluminescence (PL) is light emission from materials after the absorption of photons. Electrons are excited to higher energy levels by photons. And the excited electrons fall from the excited state to ground state. By measuring the luminescence spectrum, it is possible to observe material imperfections and impurities. By analyzing the relaxation time with time-resolved PL spectroscopy, it is possible to see the recombination process within the semiconductors and at the interfaces.

In this book, steady-state time-integrated photoluminescence (PL) measurements on the perovskite films were obtained by confocal Raman microscopy systems (Witec Alpha 300), with the excitation wavelength of 532nm. The grating of the PL spectrometer was 600g/mm. we performed a PL mapping on the perovskite films with and without Au dimers, which consists of 30,000 pixels and covers 0.15×0.15 mm^2 in one mapping.[1]

As of the time-resolved PL (TR-PL) measurements, experiments samples were excited with the Coherent Helios 532 nm nanosecond laser with a pulse width of 0.85 nanoseconds and with a repetition rate of 1 kHz. Typical pulse energies were in the range of several μJ. The PL of the samples was collected by an optical telescope (consisting of two plano-convex lenses), and it was further focused on the slit of a spectrograph (PI Spectra Pro SP2300), and eventually detected with a Streak Camera (Hamamatsu C10910) system with a temporal resolution of 1.4 ps (picosecond). The data was acquired in photon counting mode using the Streak Camera software (HPDTA).

Reference

1 Ma, C. *et al.* Plasmonic-Enhanced Light Harvesting and Perovskite Solar Cell Performance Using Au Biometric Dimers with Broadband Structural Darkness. *Sol Rrl* **3**, 1900138 (2019).

2 Raúl J.Martín-Palma, A. Vapor-Deposition Techniques. *Engineered Biomimicry*, 383-398 (2013).

3 Ma, C. *et al.* Heterostructured WS2/CH3NH3PbI3 Photoconductors with Suppressed Dark Current and Enhanced Photodetectivity. *Adv. Mater.* **28**, 3683-3689 (2016).

4 Dendooven, J. *et al.* Mobile setup for synchrotron based in situ characterization during thermal and plasma-enhanced atomic layer deposition. *Rev. Sci. Instrum.* **87**, 113905 (2016).

5 *X-ray crystallography*, <https://en.wikipedia.org/wiki/X-ray_crystallography> (2020).

6 Peng, H. *NONVOLATILE RESISTIVE RANDOM ACCESS MEMORY DEVICES: FABRICATION AND CHARACTERIZATION*, NTU, (2013).

7 *Scanning electron microscope*, <https://en.wikipedia.org/wiki/Scanning_electron_microscope> (2020).

8 *Atomic force microscopy*, <https://en.wikipedia.org/wiki/Atomic_force_microscopy> (2020).

9 *Raman spectroscopy*, <https://en.wikipedia.org/wiki/Raman_spectroscopy> (2020).

Chapter *4*

Zero-Dimensional Structural Dark Au-dimer Enhanced Perovskite Solar Cells with Plasmonic Enhanced Light Harvesting

The work presented in this chapter was published in *Solar RRL* **3**(8), 1900138.

Synopsis

This chapter demonstrates the fabrication of perovskite solar cells embedded with zero-dimensional Au-dimer. We exploited for the first time the application of nanostructures of Au nanorod-nanoparticle dimers with structural darkness to enhance the light harvesting and the performance of perovskite solar cells. Differing from conventional metallic nanoparticles, our biometric nanoparticles introduce the geometric singularity (kissing point between the nanorod and nanosphere) to the system, providing the broadband response for energy harvesting. The designed Au dimers are capable of achieving an almost ideal black-body absorption of 98–99% in a wide wavelength range between 400 and 1,400 nm. By embedding the core-shell gold dimers in the perovskite solar cells, we observed an enhancement of broad-band light absorption, and sequentially the efficiency of perovskite solar cells increases by 16%. Our results on such Au dimer embedded perovskites promote their future applications in not only solar cells, but also other optoelectronic devices.

4.1 Introduction

Photovoltaic technologies have been intensively pursued as a result of the high global demand of clean renewable energy[1-3]. Hybrid perovskites have recently attracted enormous attention for photovoltaic applications with certificated power conversion efficiency of 24.2%[3], owing to their superior physical properties such as long diffusion length[4-6], low trap density[6,7], suitable band gap and high light absorption[8-11]. Various strategies related to light management and photo-carrier collection have been developed to enhance the performance, particularly the energy conversion efficiency[1,2]. Effective light management can be accomplished by (1) reduction of reflection losses at the cell surface and (2) trapping of light in the absorbing layer[11,12]. As an effective route towards near-field light enhancement, metal nanostructures with subwavelength dimensions can couple incident photons with conduction electrons, giving rise to localized surface plasmon resonances (LSPR). However, the efficiency enhancements through the plasmonic routes are limited at short wavelength, corresponding to metal extinction wavelength[12-15]. Thus, exploration of novel plasmonic nanostructures with predesigned size and shape is needed to advance this field[16,17]. Such plasmonic nanostructures have been intensively pursued in the past decade to improve the efficiency of a wide range of solar cells, including polymer solar cells, dye-sensitized solar cells and heterojunction solar cells[1,2,18,19]. However, there have been less reports on plasmonic enhancement of the efficiency for the organometallic hybrid perovskite solar cells.

Snaith and coworkers introduced silica-coated Au nanoparticles into perovskite solar cells and achieved an efficiency enhancement of 6.5%[20]. In a later report, the same group used Ag@TiO_2 to improve the efficiency enhancement to 20% [21], which was prone to the mechanism of photon recycling. In the work by Yuan et al.,[22] an efficiency enhancement in the perovskite solar cells of ~12 % was achieved using a sandwiched TiO_x-Au-TiO_x layer as the electron transporter. The performance enhancement could be attributed to a collective effect of LSPR enhanced light absorption, reduced exciton binding energy[20], hot electron injection[22,23] and plasmonic-induced photon recycling[21]. However, so far there has been no direct experimental evidence that the enhancement of efficiency is correlated with the LSPR enhanced light absorption. Furthermore, conventional Au and Ag nanoparticles are not effective on enhancing the light absorption of perovskites , due to their high absorption coefficient (~10^5 cm^{-1}) [24-26] in the range of 500 nm - 650 nm[12,27]. On the other hand, studies have demonstrated that perovskites suffer from low light harvesting in the long wavelength range; the absorption coefficient significantly decreases in the range of 650 nm - 800 nm, and the absorption diminishes when the wavelength is longer than 800 nm[10,24,28]. Thus far, the plasmonic-induced enhancement of solar cell performance is limited due to the narrow resonance of the conventional metallic nanostructures. Traditional plasmonic structure designs usually exhibit enhancement at a certain peak or region in the wavelength, making this an ideal design for optical antennas, sensors, or light-emitting diodes. However, in order to design ideal plasmonic structures specially for photovoltaics, broadband enhancement from ultraviolet to infrared region of the solar spectrum is desired.

In this work, for the first time, we exploit a novel nanostructure of Au nanorod-nanoparticle dimers with structural darkness to enhance the light harvesting and the performance of perovskite solar cells. Differing from conventional metallic nanoparticles, these Au dimers are capable of achieving an almost ideal black-body absorption of 98–99% in a wide wavelength range between 400 and 1,400 nm, and such strong light absorption is insensitive to the angle and polarization of the incident light[29]. Besides, in such structure-engineered Au nanostructures, the sharp contact formed between the sphere and rod adiabatically focuses the incident light at all wavelength, resulting in effective enhancement in both near-field and scattering cross section, which is ideal for enhancing the performance photovoltaic devices. By embedding the core-shell gold dimers in the perovskite solar cells, we observed a notable enhancement of broad-band light absorption, and sequentially the efficiency of perovskite solar cells increases by 16%. We further investigated the photocurrent enhancement mechanism using numerical simulation, which closely matches our experimental results and guided us to optimize the position of gold dimers. Our results on such Au dimer embedded perovskites promote their future applications in not only solar cells, but also other optoelectronic devices. Furthermore, this approach of using such structure-engineered plasmonic metallic nanostructure could be generalized to other material systems to enhance the light harvesting in long-wavelength range.

4.2 Experiment Results and Discussion

Au nanorod-nanoparticle dimers were synthesized with the assistance of a thiol ligand, and a silica shell with a thickness of 2 nm was coated by the hydrolysis/condensation of tetraethyl orthosilicate (TEOS)[20]. **Figure 4.1a** illustrates a typical transmission electron microscopy (TEM) image of our Au nanorod-nanoparticle dimers. Each nano-absorber is composed of a nanorod of length 75 ± 7 nm and diameter 18 ± 2 nm and a nanosphere with a diameter of 30 ± 3 nm. As shown in the previous report[29-31], the kissing points between the spheres and the rods are formed by metallic bonding of differently curved surfaces, which have the same crystalline structure. To investigate the localized surface plasmonic resonance (LSPR) effect of the Au dimer embedded $CH_3NH_3PbI_{3-x}Cl_x$ perovskite film, a finite element method (FEM) simulation was used to calculate the field distribution in the vicinity of the Au dimer under a continuous plane wave at different wavelengths. In the simulation, the fabricated Au dimer was covered by a 2 nm-thick of SiO_2, and surrounded by perovskite. The geometrical parameters are in concordant with experiments, and the optical constants of $CH_3NH_3PbI_{3-x}Cl_x$ perovskite film are abstracted from previous studies[32,33]. **Figure 4.1b** presents the 2-D plot of the simulated near-field intensity distribution at a wavelength of 650 nm in the Au dimers embedded hybrid perovskite film. It is clear that an obvious near-field intensity enhancement has been achieved in the vicinity of the Au dimer.

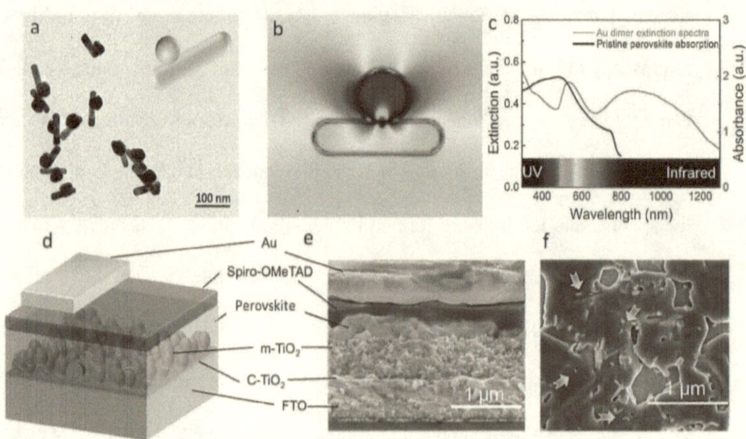

Figure 4.1(a) TEM image of Au dimers. Inset is a schematic illustration of the Au nanorod-nanoparticle dimer. (b) 2-D plot of the near-field intensity enhancement for an Au dimer. (c) Extinction UV-Vis spectra for Au dimers in water, in comparison with the absorption data of pristine perovskite. (d)(e) Schematic and SEM cross-section image of the solar cell structure. (f) SEM top view image of a perovskite film embedded with Au dimers (highlighted by arrows).

To examine the structural darkness of our Au dimers, we conducted UV-Vis experiments on the Au dimers suspension in ethanol. The extinction spectrum is plotted in **Figure 4.1c**, and the digital photo of the Au dimer solution is shown in supplementary **Figure 4.2**. In line with the previous reports[29-31], the Au dimer solution shows a broadband optical absorption, which consists of two broad peaks: one is located at 560 nm, and the other is located at 912 nm in the infrared region. The peak at 560 nm is in accord with the plasmonic resonance of Au nanoparticles[34], and the broad peak located in the infrared region (912 nm) origins from the inter-scattering of photons between nanoparticles and nanorods[17,23].

These novel Au dimer nanostructures may enhance the light harvesting of perovskite solar cells owing to several plausible mechanisms. First, leveraging on the inter-scattering of photons, these Au dimers may help extend the propagating path of photons and increase the localized light absorption. light can also be trapped in the photovoltaic cell by converting it into surface polarization polaritons (SPPs) via electromagnetic waves travelling along the interface of perovskite layer. The propagation distance can be orders of magnitude larger than the optical absorption length. Second, the sharp contact formed between the sphere and rod adiabatically focuses the incident light at all wavelength, resulting in effective enhancement in both near-field and scattering cross section. Third, the strong enhancement of the localized surface plasmon resonances (LSPR) around the plasmonic structures can be used to increase the absorption of the surrounding semiconductor materials.

Figure 4.2 Digital image of the Au-dimer in ethanol.

To investigate the effect of Au dimers on the performance of solar cells, we fabricated the conventional perovskite solar cells with mesoporous TiO_2 scaffold and compact TiO_2 as electron transport layer (ETL). **Figure 4.1d** shows a schematic of the device structure and **Figure 4.1e** shows a SEM cross-section image corresponding to the device structure. A 70 nm thick of compact TiO2 (c-TiO_2) was formed as ETL by spin coating (2000 rpm, 30 s) titanium-isopropanol solution on top of fluorine-doped tin oxide (FTO), followed by an annealing at 550 °C for 30 minutes. A 300 nm thick of mesoporous TiO_2 scaffold was then deposited on the top of the c-TiO_2, followed by a thermal treatment at 200 °C for 10 minutes and an annealing at 550 °C for 30 minutes. The perovskite infiltrated this scaffold, and a capping layer of perovskite was formed on the top, which incorporates the Au dimers serving as the active layer of the perovskite solar cells. A layer of Spiro-OMeTAD with lithium salt doping was then spin coated as a hole transport layer (HTL). Au electrodes were thermal deposited with a thickness of 100 nm.

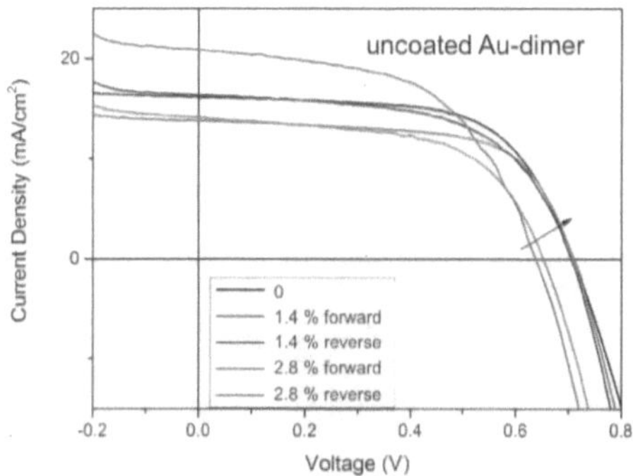

Figure 4.3 J-V curves of uncoated Au-dimer/perovskite solar cells, which show an obvious hysteresis.

The Au dimers were covered with a thin layer (2 nm) of SiO_2 shell simply to act as high refractive index dielectric conformal coating to prevent the Au dimers from the corrosion of the perovskite[35]. As supplementary **Figure 4.3** suggested, bare Au dimers would act as recombination centers in the perovskite solar cells, and Au dimers with insulating SiO_2 shells have better thermal stability[20,36]. The Au dimers@SiO_2 nanoparticles were incorporated into the perovskite layer by directly adding Au dimers DMF suspension into perovskite DMF solution at different weight ratios up to 5.6% (Au dimers@SiO_2/perovskite) while keeping the overall perovskite DMF solution with the same concentration (0.88 M in DMF) and the same thickness of perovskite capping layer[37]. At last, **Figure 4.1f** shows the top view scanning electron microscopy (SEM) image of the perovskite film incorporated the Au dimers@SiO_2 nanoparticles, demonstrating a uniform dispersion of the Au dimers@SiO_2 into the perovskite film. To further investigate the effect on the crystallinity of the Au dimers@SiO_2 embedded perovskite film, the X-ray diffraction (XRD) spectra were conducted and plotted in **Figure 4.4**. We have carefully compared the XRD spectra of the perovskite films with and without the addition of Au dimers@SiO_2. All of the perovskite films showed robust and sharp peaks at 14.10° (110), 28.41° (220), which is in line with the previous report[38,39] and indicates the influence on the crystal quality of perovskite is negligible, even with the highest addition dose up to 5.6%.

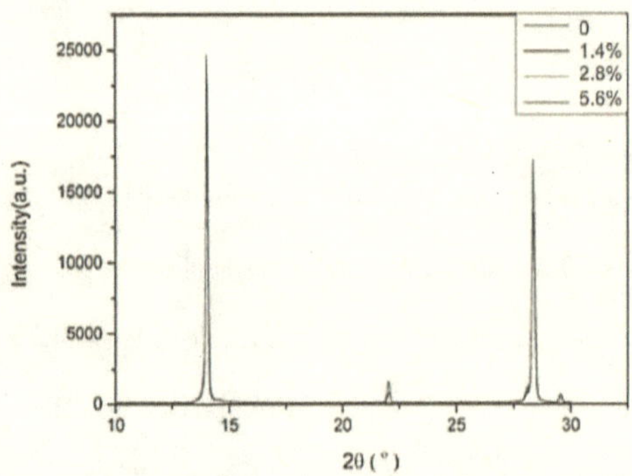

Figure 4.4 XRD pattern of the $CH_3NH_3PbI_{3-x}Cl_x$ film on glass containing Au-dimer. This shows a pure phase of perovskite and neglectable impact of the addition of perovskite on crystallinity.

Figure 4.5a shows representative photocurrent density-voltage (J-V) curves of the perovskite solar cells with different concentrations of the Au dimers, while keeping the other device parameters the same. The control device without any Au dimer exhibits a power conversion efficiency (PCE) of 14.45%, while adding Au dimers with a concentration of 1.4 wt.% increases PCE to 15.12%, mainly owing to enhanced short circuit current (J_{sc}). The highest PCE of 16.78% was achieved at an Au dimer concentration of 2.8 wt.%, which is a 16.12% enhancement with respect to the control sample. Although the fill factor (FF) slightly decreases, the increase of J_{sc} from 18.46 mA/cm^2 to 19.74 mA/cm^2 is the most significant factor behind the Au dimer-enhanced device performance. We also observed an improvement in open-circuit voltage (V_{oc}), which could be attributed to hot electron injection. The excited hot electrons tunneling across the SiO_x shell could inject into the perovskite layer and result in trap healing effect.[18] Furthermore, we observed a notable reduction of electric hysteresis in the devices with Au dimers (**Figure 4.4**). However, with further increase the Au dimer concentration of 5.6 wt.%, all solar cell parameters, *i.e.*, J_{sc}, FF and V_{oc} dramatically decreased, which may result from the dielectric shell-induced current damping as our simulation indicates (**Figure 4.6**).

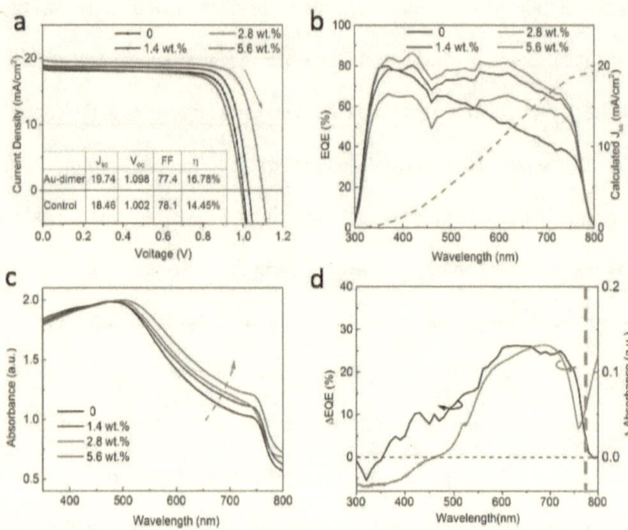

Figure 4.5 (a) Representative J-V curves of perovskite solar cells with varied concentrations of Au dimers. (b) The EQE spectra of control sample and different addition of Au dimer@SiO_2, dashed line is the integrated J_{sc} from IPCE with respect to AM1.5 solar spectrum. (c) The absorption spectra of perovskite films with vary addition of Au dimer. (d) The change of EQE (ΔEQE) and light absorption (ΔAbsorption) with the addition of Au dimer@SiO_2. The dashed line corresponds to the perovskite bandgap.

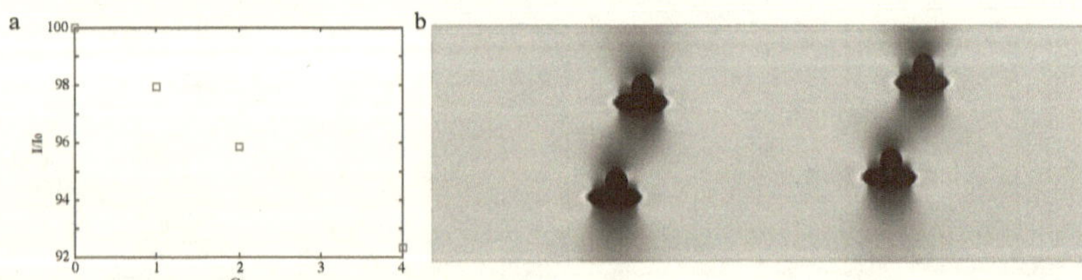

Figure 4.6 Calculated the relative current suppression by the different concentration of Au-dimers in the dark. Higher concentration of the added dimers will serve as recombination center and contribute to the series resistance.

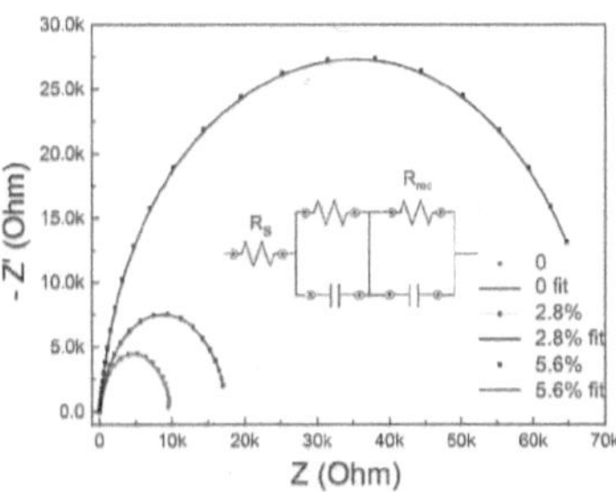

Figure 4.7 Nyquist plots of the impedance data measured on complete solar cells with different addition of Au-dimer@SiO_2. Inserted is the Equivalent circuit of perovskite solar cells used for fitting Nyquist plots.

The currents are suppressed by the addition of dielectric shell of Au dimers in the dark. The impedance spectrum measured in the dark (**Figure 4.7**) also confirmed the increasing series resistance due to the addition of the Au dimers. The fitted parameters are listed in **Table 4.1**. To ensure the reproducibility of the perovskite solar cells, we fabricated 96 devices and counted the key parameters (**Figure 4.8**). The average efficiency enhanced notably by 10.1% (from 14.11% to 14.43%) at the optimized addition concentration of 2.8%, which mainly results from the enhancement of the average J_{sc} (from 17.05 mA/cm^2 to 19.35 mA/cm^2). Therefore, it can be confirmed that the addition of Au dimers@SiO_2 has a positive effect on the performance of solar cells due to the improvement of the short circuit photocurrent.

Table 4.1 Fitted parameters of impedance spectrum for perovskite solar cells in the dark with different addition of Au-dimer@SiO_2

Addition of Au-dimer@SiO_2	0	2.8 %	5.6 %
R_s (Ohm)	14.8	15.9	26.6
R_{rec} (k Ohm)	5.49	10.1	61.2

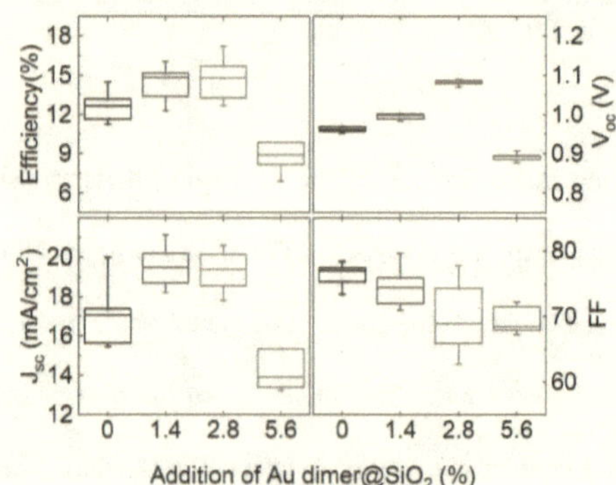

Figure 4.8 The statistical data of the performance of the Au-dimers@SiO_2 embedded perovskite solar cells

To further investigate and confirm the enhancement of photocurrent, we performed an external quantum efficiency (EQE) measurement, which refined the enhancement of photocurrent in different wavelength ranges. In **Figure 4.5b**, we show a typical EQE spectra of the control device (black line) and the ones with different additions of Au dimer@SiO_2. We observed a drop of EQE from 500 nm to 800 nm with the control device, which was also seen in previous studies [16],[18]. In agreement with the J-V curves shown in

Figure 4.5a, the increasing addition of the Au dimers lead to an enhancement EQE in the long wavelength region at first and then an entire drop at all wavelength. The dashed line in the **Figure 4.5b** shows the integrated J_{sc} from IPCE with respect to AM1.5 solar spectrum, matching well with the measured J_{sc} and confirming the validity of our measurements.

We sequentially expected the enhancement of photocurrent resulted from the enhancement of light absorption through plasmonic effect in the vicinity of Au dimers. Thus, we systematically investigated the light harvesting properties of the perovskite films with different additions of the Au dimers by ultraviolet-visible (UV-Vis) spectroscopy. **Figure 4.5c** shows the absorption spectra of perovskite films with various additions of the Au dimers. These spectra share the similar features with typical perovskite films: an absorption edge shows at around 775 nm, which is in accordance with previous reports[39]. However, we observed an obvious enhancement of light absorption in the spectral range of 475 - 800 nm, and the enhancement ratio of light absorption is proportional to the addition concentration of the Au dimer@SiO_2, consistent with the optical simulation **(Figure 4.11).**

To further demonstrate the relationship between the enhancement of light absorption and the enhancement of EQE, we calculated the increase in EQE (ΔEQE) and in absorbance (ΔAbsorbance) on the EQE spectrum and the UV-Vis spectrum with the addition of 2.8%, at which we observed the highest enhancement of solar cells efficiency (**Figure 4.5d**). Although there are some features in the spectrum, the enhancement of EQE generally coincided with the enhancement of light absorption, which further confirmed our expectation. Interestingly, both EQE and light absorption spectra illustrate a bulge in the spectral range of 500-750 nm, which indicates the increase in J_{sc} is mainly caused by the increased light absorption in this wavelength range.

To further shed light on the interaction between hybrid perovskite and Au dimers, we carried out steady-state time-integrated photoluminescence (PL) measurements on the perovskite films with and without Au dimers. We selected two representative excitation wavelengths at 473 nm and 633 nm; according to the UV-Vis absorption data shown in **Figure 4.5c**, enhancement of Au dimer-induced light absorption was only observed in the wavelength range of 600 – 700 nm. At 473 nm (2.62 eV), adding Au dimers leads to negligible change of light absorption. but we observed an obvious PL quenching, which becomes more severe with increasing Au dimer concentration.

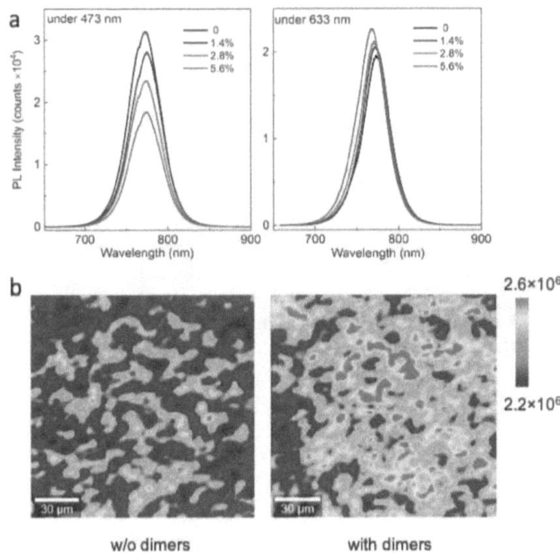

Figure 4.9 (a) Photoluminescence spectra of perovskite films with different concentrations of Au dimers, measured under light excitations of wavelengths of 473 nm laser (left) and 633 nm laser (right). (b) PL mappings at 770 nm of the two samples under the excitation of 532 nm laser.

This result shown in **Figure 4.9a** demonstrates that the presence of the Au dimers quenches the perovskite photoluminescence, indicating the either enhanced photo-recycling effect or induced charge separation, which is consistent with the previous reports[21,40]. However, we observed a reversed trend at the excitation wavelength of 633 nm (1.96 eV) in **Figure 4.9b**: the photoluminescence increased proportionally to the Au dimers addition concentration. We assumed that the increased light absorption at 633 nm by incorporated perovskite film leads to more excitons and free carriers.

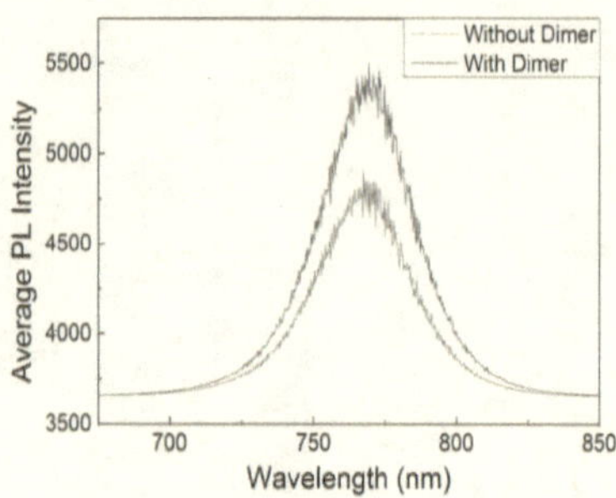

Figure 4.10 Integrated average PL intensity of perovskite film w/ and w/o Au-dimers. Due to the enhanced light absorption by plasmonic effect, the PL emission is generated in the perovskite film with Au-dimers.

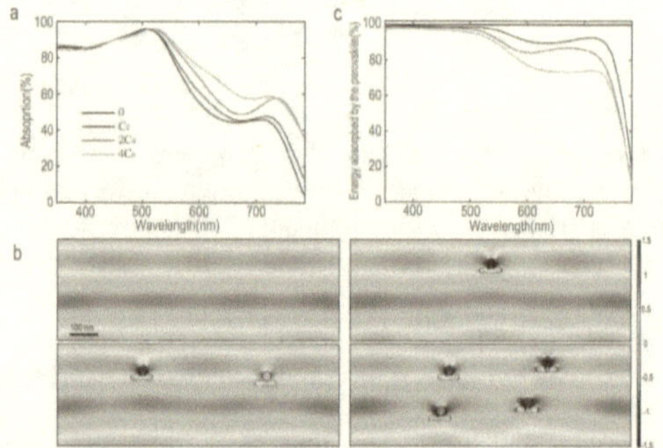

Figure 4.11 (a) The simulated absorption spectra of the perovskite film with different addition of Au dimer. (b) A simulated 2-D plot of the field distribution for different addition of Au dimer. (c) The percentage of light absorption in the perovskite film.

Considering carriers in perovskite films are dominated by free charges, electrons and holes recombine and generate photoluminescence without the separation by ETL and HTL[41]. Thus, the competitive mechanisms of enhanced photo-recycling effect and plasmonic effect induced high recombination together give rise to slight enhancement of PL intensity at the excitation of 1.96 eV. To probe the influence of the Au dimers on PL spectrum in depth, we performed a PL mapping on the perovskite films with and without Au dimers, which consists of 30,000 pixels and covers 0.15×0.15 mm^2 in one mapping. The average PL intensity is plotted in **Figure 4.10**, which shows an overall enhancement of PL intensity. The PL mapping was performed under 532 nm excitation and detected at 770 nm, illustrating an obvious PL enhancement on the perovskite film embedded by the Au dimers@SiO_2. At the same time, we also observe a blue shift and broadening of PL peak, which supposed to result from more electrons are excited to bottom of conduction band.

In order to quantitatively explain the plasmonic-enhanced performance of the device, we modeled full-wave Maxwell equation using a FEM simulator[42]. The dielectric constant of the perovskite is based on a Drude model described by the previous reference[32], and the dielectric constant of Au is from the reference[33]. **Figure 4.11a** demonstrates the absorption of the perovskite film with different Au dimer concentrations. **Figure 4.11b** illustrated the electric field distribution inside the active layer correspondingly; field enhancement is clearly observed around the dimer structure, leading to the absorption boost shown in **Figure 4.11a**, which perfectly matched the experimental result shown in **Figure 4.5c**. From the simulation, we can easily distinguish the photons absorbed by the perovskite instead of the gold nanoparticles as shown in **Figure 4.11c**. As the concentration of the gold increase, more photons will be captured inside the nanoparticles, generating the electrons that hard to be extracted. Besides, the introduction of the Au dimers with high electrical conductivity could reduce the current due to the negative impact on the carrier transportation as shown in **Figure 4.6**. As the total effect, there is an optimized concentration that provides the best performance of the solar cell, as we demonstrated in the experiments.

Plasmonic enhanced perovskite solar cells are supposed to be achieved by different mechanisms: (1) Far-field coupling of scattered light. Incident light is scattered by high albedo metal particles, reaching distances up to hundreds of nanometers (far-field), and is ultimately reabsorbed by the perovskite film. As previous work shows[21], the presence of highly polarizable nanoparticles (Ag and Au spheres) enhances the radiative decay of excitons and increases the reabsorption of emitted radiation, representing a novel photon recycling scheme. In addition, the metal particles could also capture excessive solar photons, which are not sufficiently absorbed in a certain wavelength range complementary to the absorption of perovskite. (2) Near-field coupling of light. As previous work shows[12], light scattering by nanoparticles below the incident light wavelength can be expressed as

$$C_{scat} = \frac{1}{6\pi}\left(\frac{2\pi}{\lambda}\right)^4 |\alpha|^2 \tag{4.1}$$

$$C_{abs} = \frac{2\pi}{\lambda} Im[\alpha] \tag{4.2}$$

where C_{scat} and C_{abs} are the scattering and absorption cross-section of the particles, and α is the polarizability, which can be expressed as

$$\alpha = 3V\left[\frac{\varepsilon_p/\varepsilon_m - 1}{\varepsilon_p/\varepsilon_m + 2}\right] \tag{4.3}$$

where V is the particle volume, ε_p, ε_m are the dielectric constant of the particle and perovskite, respectively. When ε_p closes to $-2\varepsilon_m$, α approaches infinity, achieving surface plasmon resonance. The strong enhancement of the localized field around the plasmonic structures can be used to boost the enhancement of the surrounding perovskite. More electron-hole pairs generated in the active layer due to the enhanced localized resonance. Therefore, the size, shape and periodical forms of the particles are crucial to the enhancement of light absorption. As shown in **Figure 4.12**, we compared the calculated light absorption spectrum in the perovskite film with Au spheres and with Au dimers. The Au dimers will rather reinforce the long-wavelength absorption of perovskite, leading to a broadband absorption. To in-depth investigate the influence of the location of the Au dimers, we compared the calculated light absorption spectrum of the perovskite with different location of Au dimer. As **Figure 4.13** indicates, locating the dimers at the interface at bottom is better.

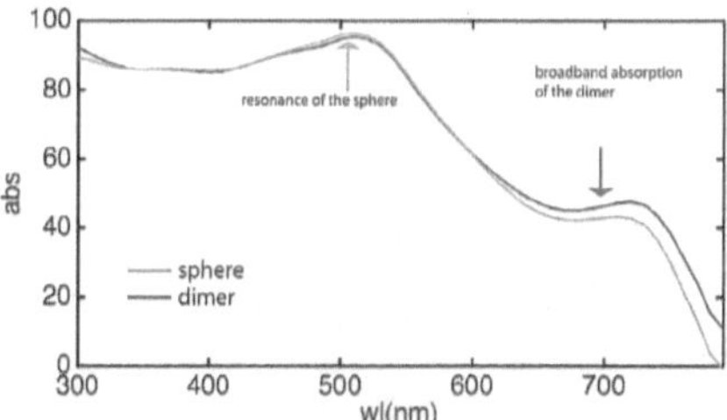

Figure 4.12 Comparison of the calculated light absorption spectrum in the perovskite film with Au spheres only and with Au-dimers. The Au dimers will reinforce the long-wavelength absorption of perovskite, leading to a broadband absorption.

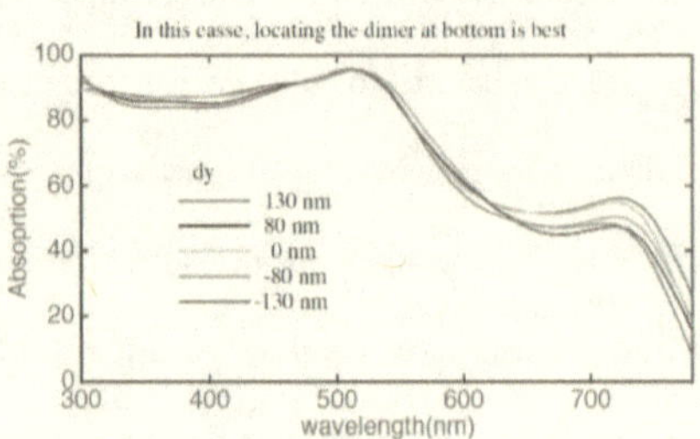

Figure 4.13 The calculated light absorption spectrum of locating the Au-dimers at different positions.

In summary, the broad light absorption enhancement in the perovskite solar cells have been achieved by incorporating the plasmonic Au dimers core-shell nanoparticles into organometal halide perovskite solar cells. Due to the broadband optical absorption enhancement in incorporated perovskite, especially in the 550-800 nm spectral range, a 19.16% enhancement on conversion efficiency has been achieved at optimized addition ratio. Plasmonic structure is widely applied in photovoltaic, however the enhancement of conversion efficiency is limited (always below 15%)[13-15,20-22,25],which results from the narrow resonance of the conventional nanostructures. Broadband enhancement is desired for the new generation of solar cells. Moreover, we systematically investigated the optical properties of the Au dimers by numerical simulation, which perfect matches our experimental results. With the superb properties of improving the light absorption in the long wavelength region, it can be expected potential for further increasing the PCE of solar cells.

4.3 Experimental Section

Material preparation: Methylammonium iodide (CH_3NH_3I) was synthesized following a previous reported method[9]. Hydroiodic acid (0.227 mol, 57 wt% in water, Aldrich) was slowly dropped into methylamine (0.273 mol, 40% in methanol) with ice-bath stirring at 0 °C for 2 h. After dropping, rotation evaporation was carried out at 50 °C for 1 h. Then the precipitate was dissolved in methanol and recrystallized by diethyl ether for three times and dried in vacuum oven for 8 h. To maintain the same thickness of perovskite films, a high concentration precursor was first synthesized. 4.6 M CH_3NH_3I and 1.2 M $PbCl_2$ (molar ratio 3:1) was dissolved in anhydrous N,N-dimethylformamide (DMF) to produce a high concentration mixed halide perovskite precursor. The high concentration precursor was diluted with Au dimers@SiO_2 (7 mg/ml) in DMF to achieve 0.8 M mixed halide perovskite precursor with different weight ratios of Au dimers@SiO_2.

Solar cell fabrication: Fluorine-doped tin oxide (FTO) was etched with Zn powder and 2M HCl and then washed with DI water. The etched substrates and bare quartz were then cleaned by detergent, DI water, and acetone, ethanol, and then dried. For optical measurement and XRD characterization, only perovskite with and without Au dimers were spin coated on the quartz substrates, thermal treated for 10 mins at 200 °C, and followed with post annealing for 1h at 550 °C. All the deposition processes are performed in glove box. For solar cells fabrication, prior to the deposition of the compact TiO_2, the FTO substrates were treated by oxygen plasma for 6 mins. A compact TiO_2 layer was spin coated (2000 rpm, 30 s) using 0.15 M titanium isopropoxide and baked at 200 °C for 2 mins, followed with a post annealing at 550 °C for 30 mins. After the substrates cooled down, a mesoporous TiO_2 layer was deposited by spin coating (2000 rpm, 45 s) diluted TiO_2 paste. The layers were sintered at 550 °C for 30 mins. The perovskite precursor solution (with and without Au dimers) was then spin coated on m-TiO_2 at 2000 rpm for 60 s and thermal treated for 1 h, which is conducted in glove box. Following previous report[4], a 68 mM spiro-OMeTAD solution in chlorobenzene containing 55 mM tert-butylpyridine and 9 mM lithium bis(trifluoromethylsyfonyl)imide (LiTFSI) salt was spin-coated the perovskite films at 2000 rpm, 45 s. Before the deposition of electrode, the cells were kept in air and in the dark for 8 h. At last, 100 nm thick of Au electrodes were evaporated by thermal evaporation.

References

1 Polman, A., Knight, M., Garnett, E. C., Ehrler, B. & Sinke, W. C. Photovoltaic materials: Present efficiencies and future challenges. Science 352, aad4424 (2016).
2 Atwater, H. A. & Polman, A. Plasmonics for improved photovoltaic devices. Nat Mater 9, 205-213 (2010).
3 Jeon, N. J. et al. A fluorene-terminated hole-transporting material for highly efficient and stable perovskite solar cells. Nat Energy 3, 682-689 (2018).
4 Sheikh, A. D. et al. Atmospheric effects on the photovoltaic performance of hybrid perovskite solar cells. Sol Energ Mat Sol C 137, 6-14 (2015).
5 Stranks, S. D. et al. Electron-Hole Diffusion Lengths Exceeding 1 Micrometer in an Organometal Trihalide Perovskite Absorber. Science 342, 341-344 (2013).
6 Xing, G. C. et al. Long-Range Balanced Electron- and Hole-Transport Lengths in Organic-Inorganic CH3NH3PbI3. Science 342, 344-347 (2013).
7 Shi, D. et al. Low trap-state density and long carrier diffusion in organolead trihalide perovskite single crystals. Science 347, 519-522 (2015).
8 Hu, X. et al. High-Performance Flexible Broadband Photodetector Based on Organolead Halide Perovskite. Adv Funct Mater 24, 7373-7380 (2014).
9 Burschka, J. et al. Sequential deposition as a route to high-performance perovskite-sensitized solar cells. Nature 499, 316-319 (2013).
10 Marchioro, A. et al. Unravelling the mechanism of photoinduced charge transfer processes in lead iodide perovskite solar cells. Nat Photonics 8, 250-255 (2014).
11 Sun, S. Y. et al. The origin of high efficiency in low-temperature solution-processable bilayer organometal halide hybrid solar cells. Energ Environ Sci 7, 399-407 (2014).
12 Wang, H. P. et al. Photon management in nanostructured solar cells. J Mater Chem C 2, 3144-3171 (2014).
13 Li, X. H. et al. Dual Plasmonic Nanostructures for High Performance Inverted Organic Solar Cells. Adv Mater 24, 3046-3052 (2012).
14 You, J. B. et al. Surface Plasmon and Scattering-Enhanced Low-Bandgap Polymer Solar Cell by a Metal Grating Back Electrode. Adv Energy Mater 2, 1203-1207 (2012).
15 Ding, I. K. et al. Plasmonic Dye-Sensitized Solar Cells. Adv Energy Mater 1, 52-57 (2011).
16 Wen, F. F. et al. Charge Transfer Plasmons: Optical Frequency Conductances and Tunable Infrared Resonances. Acs Nano 9, 6428-6435 (2015).
17 Zhang, C. et al. Al-Pd Nanodisk Heterodimers as Antenna-Reactor Photocatalysts. Nano Lett 16, 6677-6682 (2016).
18 Catchpole, K. R. & Polman, A. Plasmonic solar cells. Opt Express 16, 21793-21800 (2008).
19 Kang, Y. M. et al. Plasmonic Hot Electron Induced Structural Phase Transition in a MoS2 Monolayer. Adv Mater 26, 6467-6471 (2014).
20 Zhang, W. et al. Enhancement of Perovskite-Based Solar Cells Employing Core-Shell Metal Nanoparticles. Nano Lett 13, 4505-4510 (2013).
21 Saliba, M. et al. Plasmonic-Induced Photon Recycling in Metal Halide Perovskite Solar Cells. Adv Funct Mater 25, 5038-5046 (2015).

22 Yuan, Z. C. et al. Hot-Electron Injection in a Sandwiched TiOx-Au-TiOx Structure for High-Performance Planar Perovskite Solar Cells. Adv Energy Mater 5, 1500038 (2015).

23 DeSantis, C. J. et al. Laser-Induced Spectral Hole-Burning through a Broadband Distribution of Au Nanorods. J Phys Chem C 120, 20518-20524 (2016).

24 Barugkin, C. et al. Ultralow Absorption Coefficient and Temperature Dependence of Radiative Recombination of CH(3)NH(3)PbI(3) Perovskite from Photoluminescence. J Phys Chem Lett 6, 767-772 (2015).

25 Lu, Z. L. et al. Plasmonic-enhanced perovskite solar cells using alloy popcorn nanoparticles. Rsc Adv 5, 11175-11179 (2015).

26 Green, M. A., Ho-Baillie, A. & Snaith, H. J. The emergence of perovskite solar cells. Nat Photonics 8, 506-514 (2014).

27 Hsiao, Y. S. et al. Improving the Light Trapping Efficiency of Plasmonic Polymer Solar Cells through Photon Management. J Phys Chem C 116, 20731-20737 (2012).

28 Ma, C. et al. Heterostructured WS2/CH3NH3PbI3 Photoconductors with Suppressed Dark Current and Enhanced Photodetectivity. Adv. Mater. 28, 3683-3689 (2016).

29 Huang, J. F. et al. Harnessing structural darkness in the visible and infrared wavelengths for a new source of light. Nat Nanotechnol 11, 60-66 (2016).

30 Huang, J. F. et al. Fabricating a Homogeneously Alloyed AuAg Shell on Au Nanorods to Achieve Strong, Stable, and Tunable Surface Plasmon Resonances. Small 11, 5214-5221 (2015).

31 Huang, J. F. et al. Unravelling Thiol's Role in Directing Asymmetric Growth of Au Nanorod-Au Nanoparticle Dimers. Nano Lett 16, 617-623 (2016).

32 Loper, P. et al. Complex Refractive Index Spectra of CH3NH3PbI3 Perovskite Thin Films Determined by Spectroscopic Ellipsometry and Spectrophotometry. J Phys Chem Lett 6, 66-71 (2015).

33 Rakic, A. D., Djurisic, A. B., Elazar, J. M. & Majewski, M. L. Optical properties of metallic films for vertical-cavity optoelectronic devices. Appl Optics 37, 5271-5283 (1998).

34 Lu, L. Y., Luo, Z. Q., Xu, T. & Yu, L. P. Cooperative Plasmonic Effect of Ag and Au Nanoparticles on Enhancing Performance of Polymer Solar Cells. Nano Lett 13, 59-64 (2013).

35 Shlenskaya, N. N., Belich, N. A., Gratzel, M., Goodilin, E. A. & Tarasov, A. B. Light-induced reactivity of gold and hybrid perovskite as a new possible degradation mechanism in perovskite solar cells. J Mater Chem A 6, 1780-1786 (2018).

36 Brown, M. D. et al. Plasmonic Dye-Sensitized Solar Cells Using Core-Shell Metal-Insulator Nanoparticles. Nano Lett 11, 438-445 (2011).

37 Bera, A. et al. Perovskite Oxide SrTiO3 as an Efficient Electron Transporter for Hybrid Perovskite Solar Cells. J Phys Chem C 118, 28494-28501 (2014).

38 Tan, Z. K. et al. Bright light-emitting diodes based on organometal halide perovskite. Nat Nanotechnol 9, 687-692 (2014).

39 Li, F. et al. Ambipolar solution-processed hybrid perovskite phototransistors. Nat. Commun. 6, 8238 (2015).

40 Pazos-Outon, L. M. et al. Photon recycling in lead iodide perovskite solar cells. Science 351, 1430-1433 (2016).
41 D'Innocenzo, V. et al. Excitons versus free charges in organo-lead tri-halide perovskites. Nat. Commun. 5, 3586 (2014).
42 Polstyanko, S. V., DyczijEdlinger, R. & Lee, J. F. Full wave vector Maxwell equation simulation of nonlinear self-focusing effects in three spatial dimensions. Ieee T Magn 33, 1780-1783 (1997).

Chapter 5

Two-Dimensional Transition Metal Dichalcogenide WS_2 with $CH_3NH_3PbI_3$ Photoconductor with Suppressed Dark Current and Enhanced Photodetectivity

Synopsis

This chapter demonstrates a photoconductor (a kind of optoelectronic device) made of methylammonium lead triiodide ($CH_3NH_3PbI_3$) perovskite films and WS_2 monolayers. As a result of the interfacial charge transfer, the dark current of these devices was significantly suppressed, accompanied by remarkable photoluminescence (PL) quenching. Furthermore, the photocurrent was enhanced by about one order of magnitude, which can be attributed to the suppressed charge recombination. Both factors lead to a photodetectivity of $\sim 10^{12}$ Jones measured in photoconductors based on perovskite/WS_2 hybrid. Because the photocarriers are effectively separated in the bilayers, and WS_2 can assist electrons to transport across the channel, the response time of the photodetectors is on the millisecond level, which is much faster than the single-layer counterparts. Finally, from the structural point of view, $CH_3NH_3PbI_3$ perovskite layers grown on WS_2 monolayers exhibit improved crystallinity, which contributes to the high-performance of the hybrid WS_2/perovskite photodetectors. These results make heterostructures of perovskites and TMDs a promising candidate for constructing high-performance photodetectors and other optoelectronic devices.

5.1 Introduction

Recently, methylammonium lead halide perovskites (e.g. $CH_3NH_3PbI_3$) have attracted enormous attention due to its remarkable physical properties, such as broadband light absorption [1-4], long diffusion length and low trapping density.[5-7] These materials have triggered the revolution of a wide range of photonic and photovoltaic devices, including solar cells,[8-10] light-emitting devices,[11-13] lasers,[14-16] and photodetectors.[1,17-22] Meanwhile, advents of synthetic routes have enabled the production of perovskite films with optimal performance, low cost, high reliability and stability.[8,23-27] In optoelectronics, photodetectors are fundamental devices, enabling the conversion of light signal to electronic one and being widely applied in digital photography and cinematography.[28] So far, a wide range of materials, including PbS quantum dot,[29] GaN,[30] graphene,[31] transition metal dichalcogenides,[32] have been exploited for photodetector applications. As a result of their extraordinary physical properties, perovskites have been used as the active materials in photodetectors. So far, multilayer solar-cell-like heterostructures have presented the best photodetection capability,[17] but their structure is complex and interfaces must be carefully controlled to achieve the optimal performance. Planar-structured photodetectors are much easier to fabricate, and their simple structure helps elucidate the operating mechanism.

However, such photoconductors often suffer from large electrical hysteresis, slow photoresponse and lower sensitivity. Several efforts have been made towards overcoming these drawbacks via, for examples, fabricating hybrid bilayer photodetectors,[19] perovskite passivation,[20] phototransistor gating, etc.[33] One particularly powerful approach to improve the performance of perovskite photodetectors is to interface perovskite layers with other functional materials. For example, a perovskite/graphene heterostructure has been proposed to achieve high photo-responsivity.[19] However, because graphene does not have a band gap, the structure is limited by high Off currents, leading to low sensitivity.

Transition-metal dichalcogenides (TMDs) possess tunable band gaps[34] and high mobility,[35] which may help suppress the Off current and promote charge separation in photodetectors. Monolayered TMDs have been intensively exploited in the past decade for a wide range of potential applications, including optoelectronics,[36,37] biosensors[38] and piezoelectricity.[35] Graphene/MoS_2 heterostructures have been shown to deliver enhanced photoresponsivity compared with their constituents.[39] Among well-investigated TMDs, WS_2 (conductive band edge at -4.0 eV and valence band edge at -5.96 eV) has good band alignment[38] with the $CH_3NH_3PbI_3$ perovskite (conductive band edge at -3.75 eV and valence band edge at -5.35 eV). However, WS_2 monolayers have not been integrated with perovskite to improve the photodetection capability in planar photodetectors.

In this work, we fabricated and characterized for the first time a photoconductor made of methylammonium lead triiodide ($CH_3NH_3PbI_3$) perovskite films and WS_2 monolayers. Because of the interfacial charge transfer, the dark current of these devices was significantly suppressed, accompanied by remarkable photoluminescence (PL) quenching. Furthermore, the photocurrent was enhanced by about one order of magnitude, which can be attributed to the suppressed charge recombination. Both factors lead to a photodetectivity of ~10^{12} Jones measured in photoconductors based on perovskite/WS_2 hybrid. Because the photocarriers are effectively separated in the bilayers, and WS_2 can assist electrons to transport across the channel, the response time of the photodetectors is on the millisecond level, which is much faster than the single-layer counterparts. Finally, from the structural point of view, $CH_3NH_3PbI_3$ perovskite layers grown on WS_2 monolayers exhibit improved crystallinity, which contributes to the high-performance of the hybrid WS_2/perovskite photodetectors. These results make heterostructures of perovskites and TMDs a promising candidate for constructing high-performance photodetectors and other optoelectronic devices.

5.2 Results and Discussion

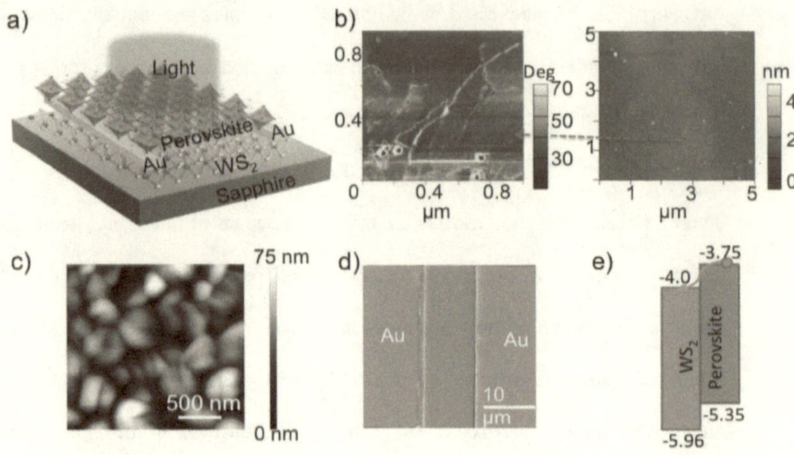

Figure 5.1 a) Schematic device structure of the hybrid WS_2/perovskite photoconductor fabricated on a C-plane (0001) sapphire substrate. The crystal structures of $CH_3NH_3PbI_3$ perovskite and WS_2 are also shown. **b)** AFM height image of the monolayer WS_2 film. Inset is a zoom-in phase image, highlighting the grain boundary. **c)** AFM Height image of the perovskite film grown upon the WS_2 layer. **d)** SEM image of the bilayer photoconductor device. The channel width is 10 μm. **e)** Band alignment of the WS_2/perovskite hybrid bilayer. Electron transfer at the interface is also sketched.

Figure 5.1a presents a schematic device structure of the hybrid WS_2/perovskite photodetector fabricated on a C-plane (0001) sapphire substrate. Experimental details are given in experiment section. Continuous polycrystalline WS_2 monolayers were grown on C-plane sapphire substrates by CVD method, and then Au electrodes with a thickness of 80 nm were fabricated on the WS_2 monolayers using photolithography and sputtering. With part of the electrodes being protected by tapes, one layer of PbI_2 film with a thickness of 150 nm was deposited by thermal evaporation. To convert the PbI_2 into $CH_3NH_3PbI_3$

perovskite, the PbI2 film was exposed to CH_3NH_3I vapors in an atomosphere-controlled vacuum chamber. After the conversion, the thickness of $CH_3NH_3PbI_3$ perovskite film reached about 300 nm, which is similar to the thickness commonly used in photovoltaic devices. It should be noted that the Au electrodes contacted both WS_2 and perovskite, thus capable of collecting charges from both layers. **Figure 5.2** shows the digital image of the photodetectors with and without WS_2. And **Figure 5.3** shows SEM image of the photodetector devices.

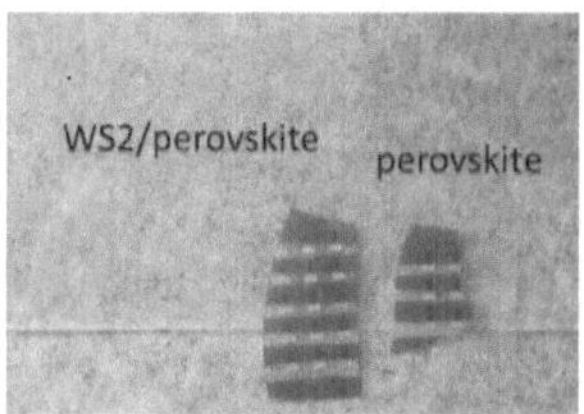

Figure 5.2 Photographs of the photoconductor devices. Right is WS2/perovskite device, and left is pristine perovskite device.

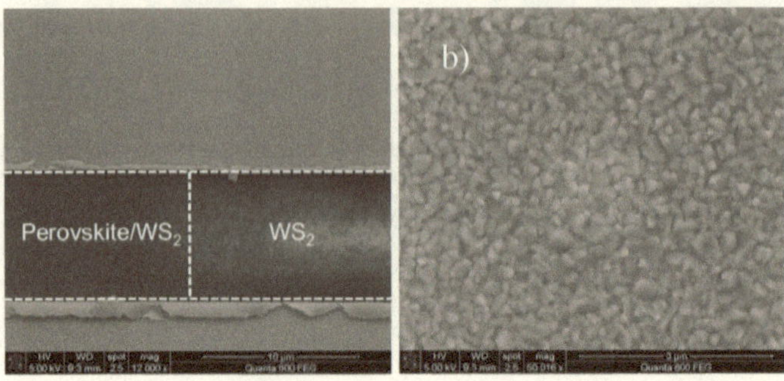

Figure 5.3 a), SEM image of the photodetector devices: left is the SEM image of perovskite/WS2 bilayer and right is the perovskite; b), enlarged SEM image of perovskite on WS2.

As illustrated in **Figure 5.1a**, $CH_3NH_3PbI_3$ has a three-dimensional distorted perovskite crystal structure belonging to the tetragonal *I4/mcm* space group, while WS_2 monolayer has a two-dimensional hexagonal lattice structure. **Figure 5.1b** shows the height image of WS_2 film measured by atomic force microscopy (AFM), and the roughness of WS_2 film is less than 1 nm. Grain boundaries can be clearly observed from the AFM phase image (**Figure 5.1b**), which is in good agreement with a previous report. [33] **Figure 5.1c** shows the AFM height image of the polycrystalline $CH_3NH_3PbI_3$ perovskite film grown on the WS_2 monolayer. The average grain size is proximately 0.3 μm, and the root-mean-square roughness is 20 nm in a typical scanning area of 5 μm×5 μm. The maximum height is approximately 84 nm, which is much less than the film thickness (300 nm), indicating that the sequential vapor prepared $CH_3NH_3PbI_3$ perovskite film provides continuous surface coverage.

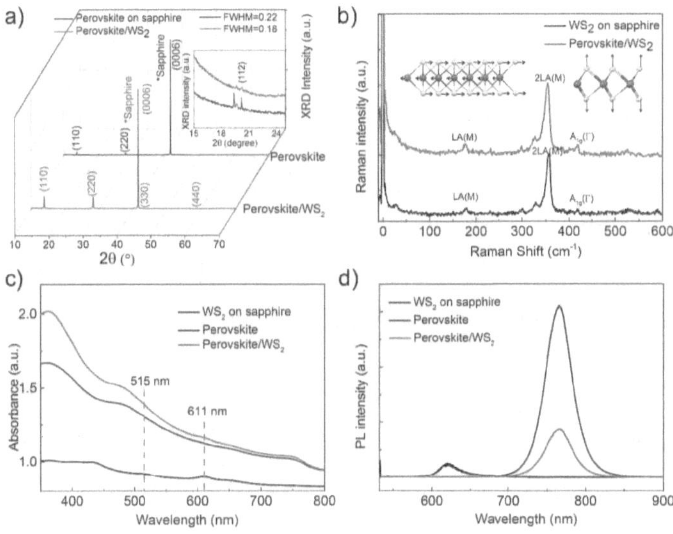

Figure 5.4 a) XRD spectra of the as-prepared $CH_3NH_3PbI_3$ perovskite samples on sapphire (blue) and on WS_2/sapphire (red). Inset highlights the (112) peak measured on the $CH_3NH_3PbI_3$ grown without WS_2. **b)** Raman spectra of the WS_2 film and the WS_2/perovskite bilayer measured under 532 nm laser excitation. Inset vibration modes represent the modes corresponding to LA(M) and $A_{1g}(\Gamma)$. **c)** Ultraviolet-visible (UV-Vis) absorption spectra for the pristine WS_2, perovskite and heterostructured WS_2/perovskite samples. **d)** Corresponding PL spectra of the samples.

To further investigate the structure of the perovskite layer deposited on the WS_2 monolayer, X-ray diffraction (XRD) measurement was conducted. **Figure 5.4a** shows the XRD patterns of the perovskite samples on sapphire (blue line) and WS_2/sapphire (red line). Peaks at 14.08°, 28.41°, 43.19°, and 58.79° can be assigned to the (110), (220), (330), and (440) crystallographic planes of $CH_3NH_3PbI_3$, respectively.[1] The absence of the PbI_2-related peak at 12.8° confirms the phase purity of the perovskite layer. In the XRD pattern of the $CH_3NH_3PbI_3$ grown without WS_2, we observed a weak but distinguishable peak at 20.6° (inset of **Fig. 5.4(a)**), which can be assigned to the (112) plane of perovskite. In contrast, this peak disappeared when the $CH_3NH_3PbI_3$ layer was grown on WS_2, indicating more coherent grain orientation. Furthermore, on WS_2, the intensity of all XRD peaks are significantly enhanced and the full width at half of maximum (FWHM) of the (110) perovskite peak was reduced from 0.22° to 0.18° (**Figure 5.5**), confirming the improved crystallinity of $CH_3NH_3PbI_3$ perovskite in the bilayers.

Raman spectra could provide valuable insights on the physical properties of WS_2, particularly on identifying the number of WS_2 layers.[40] As presented in **Figure 5.4b**, the resonance longitudinal acoustic phonon mode (2LA(M)),[41] appears much stronger than the $A_{1g}(\Gamma)$ mode, which is consistent with the high quality of the WS_2 monolayer. The vibration modes of both Raman peaks are presented inset **Figure 5.4b**. Furthermore, this feature was retained after the deposition of the perovskite layer, indicating that the deposition of PbI_2 and the conversion to the $CH_3NH_3PbI_3$ phase generated negligible defects in the WS_2 monolayer.

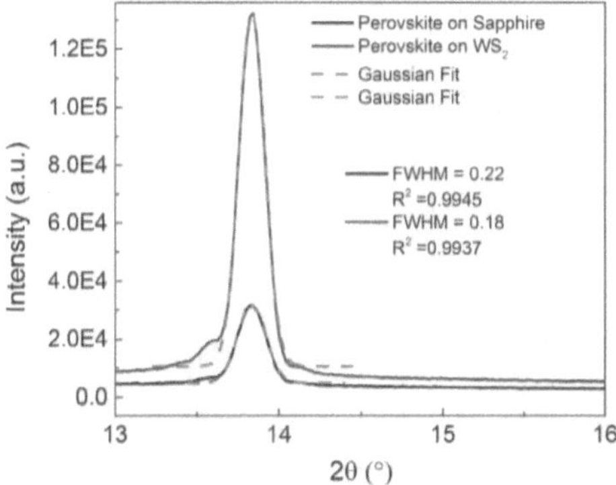

Figure 5.5 XRD patterns of different samples. Red dash lines indicate a good Gaussian fit of both peaks.

Ultraviolet-visible (UV-Vis) absorption data are presented in **Figure 5.4c**. Two absorption peaks (at 515 nm and 611 nm) were clearly observed on the pristine WS_2 film, which are related to the excitonic absorptions of the direct gap located at the K valley of the Brillouin zone, and the separation between these two peak results from the splitting of the valence band minimum due to the spin-orbit coupling.[41] After the perovskite deposition, the 611 nm peak of WS_2 is still visible. **Figure 5.4d** shows the photoluminescence (PL) spectra for the samples measured under 532 nm laser. The pristine WS_2 film exhibits a PL peak at 624 nm, which is related to the band gap of WS_2.[41] Both perovskite and hybrid samples exhibit a peak at 765 nm, which is consistent with the band gap of perovskite.[42] More importantly, a significant PL quenching was observed on the WS_2/perovskite bilayer, indicating efficient charge transfer at the interface and exciton dissociation.

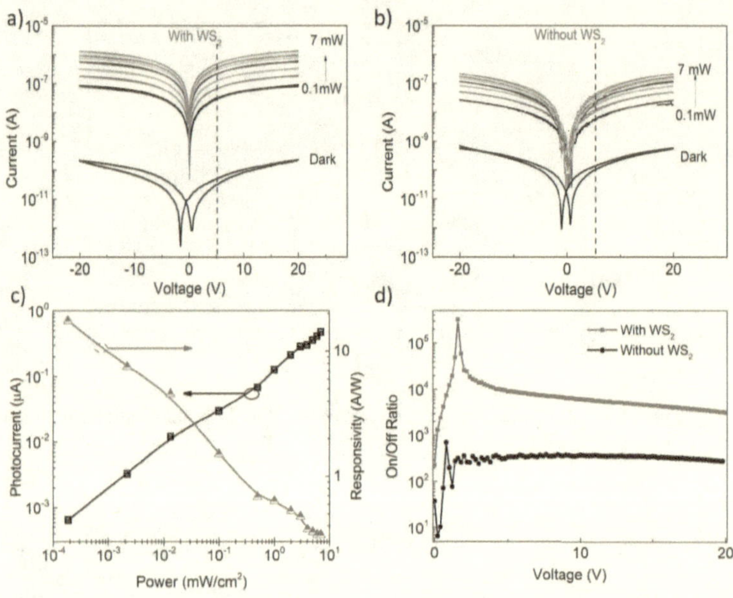

Figure 5.6 a) IV curves of the single-layer perovskite photoconductor measured in dark and under illumination with different white light intensities. **b)** Corresponding IV curves measured on the hybrid WS_2/perovskite bilayers. **c)** Photocurrent and photo-responsivity of the bilayer photoconductor measured at 5 V bias. **d)** Bias dependence of ON/OFF ratios measured on the two samples.

Due to the dissociation of photo-induced excitons and electrons transfer from perovskite to WS_2, a photoresponse enhancement can be expected in the heterostructured WS_2/perovskite. IV curves of the intrinsic perovskite measured in dark and under white light irradiation with various powers (0.1 mW/cm^2, 0.5 mW/cm^2, 1.0 mW/cm^2, 2.0 mW/cm^2, 3.0 mW/cm^2, 4.0 mW/cm^2, 5.0 mW/cm^2, 6.0 mW/cm^2, and 7.0 mW/cm^2) are shown in **Figure 5.6a**. Higher irradiation power generated higher current at the measured power range. However, when the irradiation power increased gradually, the photocurrent ($I_{ph} = I_{illuminated} - I_{dark}$) tended to be saturated. Photoconductor performances of perovskite layers with different thicknesses were presented in **Figure 5.7**.

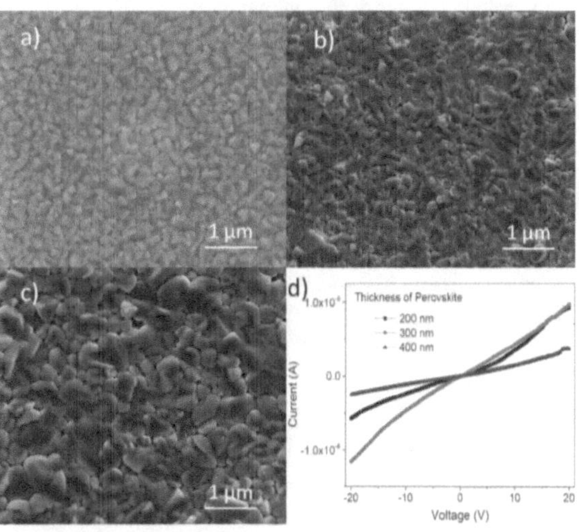

Figure 5.7 SEM images of perovskite on sapphire with different thickness: a) 200 nm, b) 300 nm and c) 400 nm, respectively. d) Photoresponse of the devices with different thickness.

By interfacing the perovskite with the WS_2 monolayer, the photocurrent was enhanced by more than one order of magnitude (**Figure 5.6b**). Under light, the linear and symmetric IV curves of the perovskite/WS_2 bilayer indicate ohmic-like contacts between the photoconductors and the electrodes, whereas nonlinear IV curves with lower photocurrents were observed in perovskite-only device (**Figure 5.8**). A more smooth, stable and Ohmic-like IV curve was observed in the hybrid system, which indicates the suppressed trapping/detrapping process at interfaces. By comparing the dark current of two systems, although they are both Schottky-like curve, a suppressed dark current (more than one order of decrease) was observed in hybrid system, which may result from the lower carrier concentration in the depletion region.

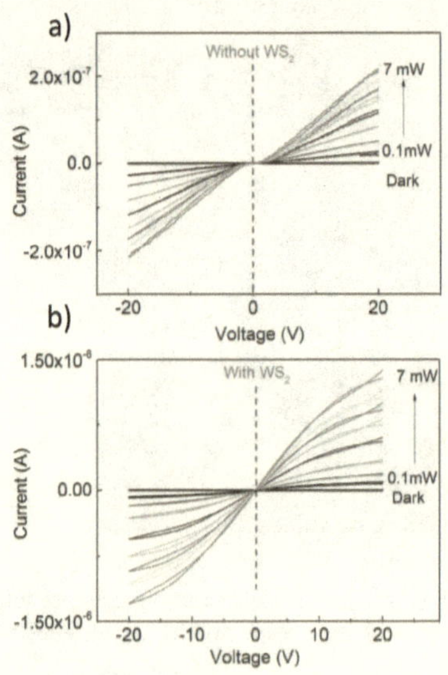

Figure 5.8 a) Linear IV curves of the single-layer perovskite photoconductor measured in dark and under illumination with different light intensities. b) Corresponding IV curves measured on the hybrid WS_2/perovskite bilayers.

Figure 5.6c shows the photocurrent and photo-responsivity when the device operates at the bias of 5 V. At light intensity of 0.2 μW/cm², the hybrid photoconductor device shows as high as 17 A/W responsivity. The photocurrent increased gradually as the irradiation power increased. Meanwhile, responsivity decreases as the irradiation power increasing, and this is consistent with the previously reported photogating mechanism.[18] Assuming the relation between responsivity and irradiation power follows $R \propto P^{-1}$, when the power is set down to 1 pW, the responsivity in our case is expected to reach ~10^3 A/W. **Figure 5.6d** presents the on-off ratio with respect to the operating bias voltage. At the bias of 1.6 V, on-off ratio of the hybrid device reaches the highest peak of 3×10^5, which results from the transition of Schottky contact in dark to Ohmic contact under light. However, the dark current is so low that reaches the equipment limit, and at this time, photoresponsivity is also low. Thus, we chose 5 V as the operating bias of the photoconductors.

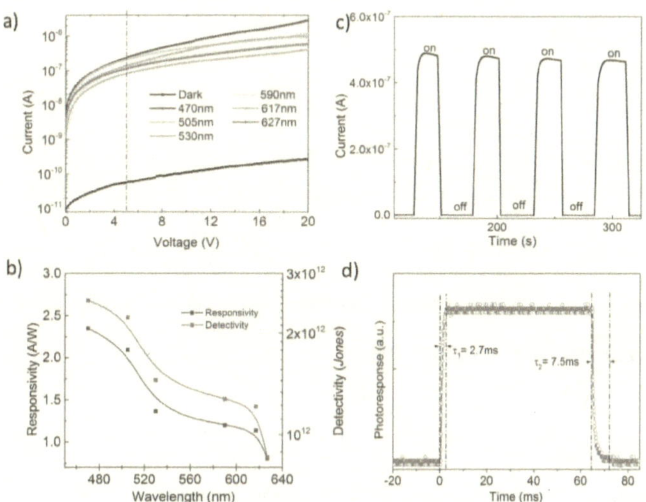

Figure 5.9 a) Wavelength dependent performance of the heterostructured Perovskite/WS_2 photoconductor under the same irradiation power 0.5mW/cm^2. **b)** Wavelength dependent responsivity and detectivity of the heterostructured WS_2/perovskite photoconductor. **c)** Time dependent current response of the heterostructured WS_2/perovskite photoconductor

under chopped irradiation (under a bias of voltage of 5V). **d)** Temporal voltage responses highlighting a rise time of 2.7 ms and a decay time of 7.5 ms, which measured by oscilloscope.

To further investigate the photoresponse of the heterostructured photoconductor at the different light wavelength, current-voltage measurement was carried out under the controlled irradiation power (0.5 mW/cm2). Wavelength dependent photoresponse was presented in **Figure 5.9a**, the corresponding photoresponsivity and detectivity at different wavelength were calculated and presented in **Figure 5.9b**, in which, responsivity and detectivity were defined as[12]

$$R = \frac{I_{illuminated} - I_{dark}}{P_{in} \times A} \tag{5.2}$$

$$D^* = \frac{(A\Delta f)^{1/2} R}{i_{total}} \tag{5.3}$$

in which, $I_{illuminated}$ is the current under illumination, I_{dark} is the dark current, P_{in} is the incident irradiation power, A presents the active area (10 μm×2 mm), Δf is the bandwidth, and i_{total} is the noise current. Because, in our devices, the noise current was dominated by the shot noise (**Figure 5.10**),

$$i_{total} \approx i_{n,s} = \sqrt{2qI_d\Delta f} \tag{5.4}$$

detectivity can be presented as the following formula.[21]

$$D^* = \frac{R}{(2qI_d/A)^{1/2}} \tag{5.5}$$

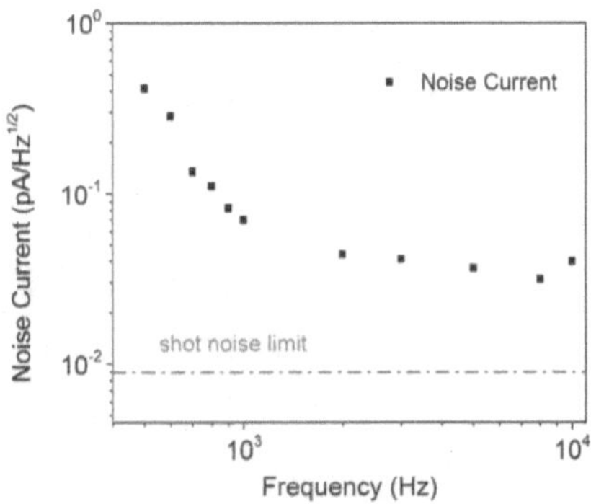

Figure 5.10 Noise current data measured by lock-in amplifier with different reference frequency, demonstrating a relative low noise current dominated by shot noise. Red dash line shows the shot noise.

Owe to its suppressed dark current and enhanced responsivity, detectivity of the heterostructure device is as high as 2×10^{12} Jones at the illumination intensity of 0.5 mW/cm^2 and the wavelength of 505 nm. As a previous report [18] and our experiment data, both responsivity and detectivity are strongly related to irradiation power. The fast-electronic response to the optical signal is crucial to all photodetectors, which is related to the charge transport and fast collection. In our heterostructure photoconductor, the thickness of perovskite was optimized to be 300 nm (plotted in **Figure 5.7**). The temporal photoresponse of the heterostructure WS_2/perovskite photoconductor was carried out under a bias of 5 V, and the irradiation power of 5 mW/cm^2 (**Figure 5.9c**), the on-off switching was reproduced for multiple cycles, which illustrates a good reproducibility of the photoconductor. For comparing, the ON/OFF switching curve of the intrinsic perovskite photoconductor was presented in **Figure 5.11**, which shows a very slow response time.

Owe to the efficient charge separation and surface passivation, the heterostructure WS_2/perovskite photoconductor shows a much quicker response.

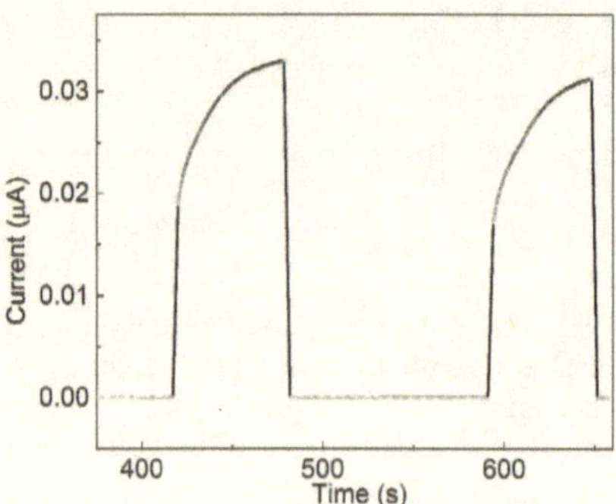

Figure 5.11 Time-dependent photocurrent response of the intrinsic perovskite photoconductor under chopped light irradiation (bias: 5 V).

Furthermore, in the present of depletion region near the WS_2/perovskite interface, the transport of electrons and holes are separated, which limits the recombination of electron and holes and promotes the response performance. To determine the accurate response time of the hybrid photoconductor, oscilloscope measurement was carried out (**Figure 5.9d**). The rise and decay times were measured to be 2.7 ms and 7.5 ms, respectively. Comparing with the intrinsic perovskite photoconductor, the heterostructure photoconductor shortens the response time by five orders. And comparing with the previous reported graphene/perovskite hybrid photodetector, the response time in our case is also shorter by three orders. As a key Figure-of-merit of detection performance,[12] the fast response broadens the application of devices, furthermore, planar structure is more facile to fabricate and accessible for micro-devices array.

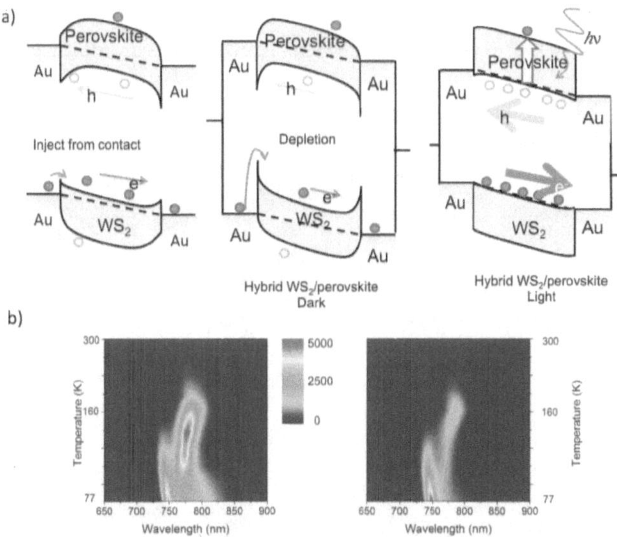

Figure 5.12 a) Proposed working mechanism of the perovskite/WS_2 bilayer photoconductor, forming a back-to-back Schottky diode, in which the perovskite and the WS_2 layers transport photo-carriers as parallel channels. **b)** Temperature dependent PL data of the intrinsic perovskite (left), and the hybrid WS_2/perovskite (right) samples.

As proposed in **Figure 5.1e**, band-alignment between perovskite and WS_2 generates the built-in field, which separates the photo-generated electrons and holes without applying any gate voltage. At the same time, electrons and holes are transported in WS_2 film and perovskite film, respectively, which accelerates the electron transporting speed by the high electron mobility of WS_2, and inhibits the recombination of electrons and holes by the built-in field. The energy structure of the back-to-back Au-WS_2 (Perovskite)-Au interface under the bias was proposed in **Figure 5.12a**.[43-45] As showing in **Figure 5.13**, pristine WS_2 film shows little light response and stays at off-state for the reason of lack of gate voltage, the Fermi level of WS_2 is lower than its bottom of conductive band (CB).[1,46]

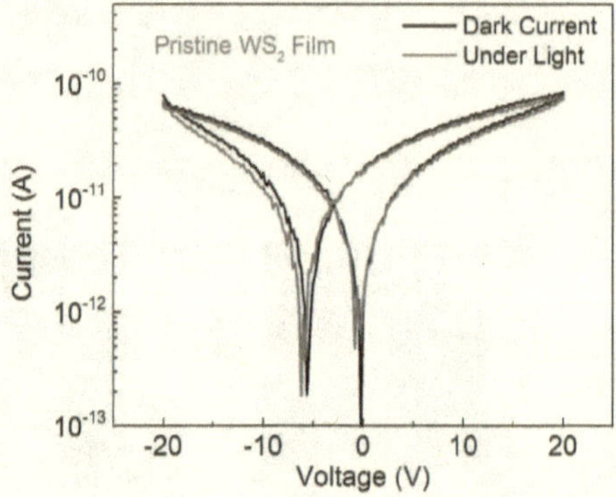

Figure 5.13 Dark curve (black) and light curve (red) of pristine WS_2 film, indicating the pristine WS_2 film stays at off state.

When the light is off, Au/WS_2 (Au/perovskite) interface induced Schottky junction hampers the tunneling of electrons (holes), whilst WS_2/perovskite interface induced depletion region lowers (raises) the Fermi level of WS_2 (perovskite), which suppresses the dark current of the hybrid system. However, when the light is on, electron-hole pairs generate in perovskite, and electrons transfer to WS_2 film, which raises the Fermi energy of WS_2, and decreases the effective height and width of Schottky barrier, resulting in easier carriers tunneling and carrier transport.

On the other hand, $CH_3NH_3PbI_3$ perovskite photovoltaic devices were studied to be free charge and excitons coexisted devices, with the ~55meV exciton binding energy in Photovoltaic (PV) regime at room temperature.[47] Both free charge carriers and excitons are generated in perovskite under light. Free charges are separated to be electrons and holes, and transported in WS_2 and perovskite, respectively. The accumulation of space charges at WS_2/perovskite interface together with band bending forms a p-n junction, as well as a depletion region near the interface of WS_2/perovskite, which assists the free charge separation. To verify the effect on excitons dissociation, temperature-dependent photoluminescence was carried out (**Figure 5.12b**). The temperature was controlled ranging from 77 K to 300 K. The left figure shows the intrinsic perovskite PL spectra at the different temperature, which is consistent with previous report,[47,48] showing a phase transition at 160K. At low temperature (below ~100K), orthorhombic PL peak (at ~750 nm) dominates, while tetragonal peak (longer wavelength) co-exists in the system.[39] The theoretical calculation also reveals the orthorhombic phase has a larger band-gap than tetragonal phase in perovskite.[49] Comparing with the left and right Figures in **Figure 5.12b**, orthorhombic peak has little difference at low temperature, while tetragonal peak quenched as that in room temperature, which indicates only tetragonal peak is related to the proper conductive band (CB) position, and contributes to the PL quenching. Since orthorhombic phase is not stable at room temperature, exciton binding energy of perovskite can be estimated from the tetragonal peak above 160 K. FE emission intensity can be fitted by[50]

$$I(T) = \frac{I_0}{1+Ae^{-E_B/k_BT}} \tag{5.6}$$

in which I_0 is the PL intensity at 0K, k_B represents the Boltzmann constant. By fitting the intensities at the different temperature, the excitons binding energy E_B can be estimated. From the fitting curve, plotted in Supplementary Information **Figure 5.14** and **5.15**, exciton binding energies for intrinsic perovskite and hybrid WS_2/perovskite were estimated to be 59.0±8.5 meV, and 30.5±8.3 meV, respectively. A dramatic decrease of exciton binding energy was observed, which indicates WS_2/perovskite induced built-in barrier assists the dissociation of excitons in perovskite.

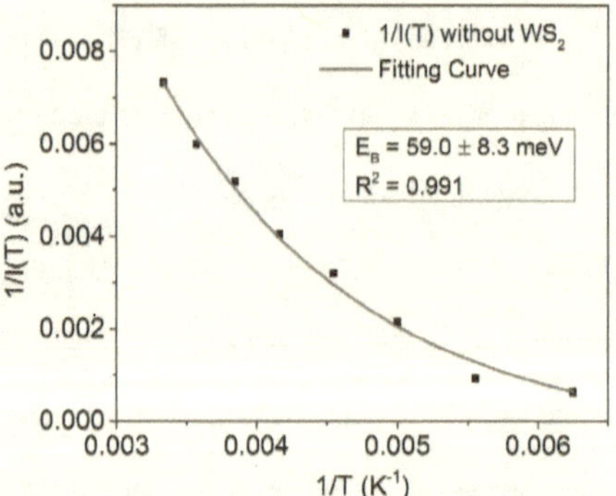

Figure 5.14 The intensity of temperature dependent PL of the sample without WS_2, red solid line illustrates the fitting curve of the PL intensity.

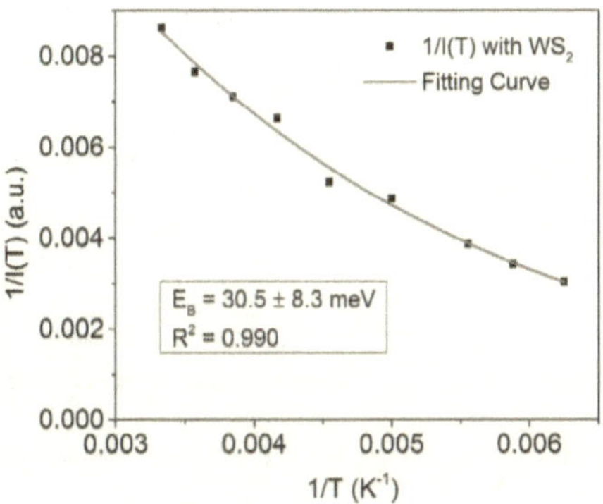

Figure 5.15 The intensity of temperature dependent PL of the sample with WS2, red solid line illustrates the fitting curve of the PL intensity.

In this work, high-performance photoconductor based on the hybrid perovskite and 2D materials is fabricated and characterized for the first time. Due to the superior properties of both perovskite and WS_2, photoconductor in this work shows the highest performance on detectivity among all the planar structure perovskite photodetector. Remarkably, the photodetectors exhibit more reliable high performance on ON/OFF ratios (~10^5) and high responsivity (~17 A/W) under 10^{-4} mW/cm^2 irradiation power. Thanks to the high mobility of WS_2 film and the charges transport separation by built-in barrier, the response time of hybrid WS_2/perovskite photoconductors are enhanced by four orders (from tens of seconds to milliseconds). Furthermore, WS_2 film contributes as an atomically flat and highly ordered substrate, perovskite film grows on it shows better orientation and better crystalline properties, which are benefit to the electrons transport and result in a better charge transfer.

According to previous report,[51] WS_2 film can also be a superior passivation layer, which is benefit to the charge release from the interface, and then improves the response properties of photodetectors. At last, WS_2/perovskite induced built-in barrier provides a depletion region to inhabit the recombination of electron and holes, which lead to a better photoresponse and faster response speed. Our results for the first time built the high-performance TMD/perovskite heterojunction photoconductors, and provide a useful insight into charge transport in hybrid WS_2/perovskite system, promoting their future applications in optoelectronic devices.

5.3 Experimental Section

Material Synthesis and Device Fabrication: The WS_2 monolayer films were synthesized on clean C-plane sapphire substrates using chemical vapor deposition (CVD) method, where WO_3 powder (300 mg, Sigma-Aldrich) in a quartz boat was put in the heating zone at the center of a tube furnace. The sapphire growth substrates were placed at the downstream next to WO_3 powders. S powers were put in a separated quartz boat at the upper stream where the heating temperature can be independently controlled. In a typical CVD process to grow WS2 monolayers, Ar/H_2 gas flow (Ar = 50 sccm, H_2 = 10 sccm) and pressure of 5 Torr were used. To start the growth, the heating zone for WO_3 was raised to 900 °C in 30 mins, and the temperature for S was raised to 240 °C. After the furnace tube reached the growth temperature, the growth lasted for 15 mins, which was followed by a natural cooling to room temperature.

The $CH_3NH_3PbI_3$ perovskite films were deposited onto the WS_2 monolayers using the sequential vapor deposition method, following the previous report.[26] As-prepared WS_2 monolayer films were loaded into a thermal evaporation chamber (base pressure is below 1×10^{-4} Pa) to deposit PbI_2 films. The evaporation temperature is 152 °C and the evaporation rate was 0.4 Å/s. After the PbI_2 film thickness reached 150 nm, samples were put on a hot plate in an atmosphere-controlled chamber, with CH_3NH_3I powder surrounding the substrate, and the PbI_2 films were converted into $CH_3NH_3PbI_3$ at a temperature of 120 °C. After two hours of phase conversion, one hour of post-annealing was carried out with DMF vapor to passivate the perovskite surface. CH_3NH_3I powder was synthesized following the previous report:[2] hydroiodic acid (0.227 mol, 57 wt% in water, Aldrich) was slowly dropped into methylamine (0.273 mol, 40% in methanol) with ice-bath stirred at 0 °C for two hours. After dropping, rotation evaporation was carried out at 50 °C for one hour to produce CH_3NH_3I. The precipitate was dissolved in methanol and recrystallized by diethyl ether for three times and dried in vacuum oven for eight hours.Au electrodes (80 nm) were fabricated using photolithographed and sputtering, which defined a channel length (L) of 2000 μm and channel width (W) of 10 μm.

Measurements: All the devices and films were measured and characterized in ambient condition but stored in vacuum desiccator. Perovskite and PbI_2 film thickness was assessed by atomic force microscopy (AFM, Bruker Dimension ICON), and monolayered WS_2 film surface was assessed by atomic force microscopy (AFM, Veeco Dimension-Icon system). Both AFM were conducted at the tapping mode. The photoluminescence (PL) spectra and the Raman spectra of pristine WS_2, WS_2/perovskite hybrid sample, and perovskite films on sapphire substrate were obtained by confocal Raman microscopy systems (Witec Alpha 300), with the excitation wavelength of 532nm. The grating of PL spectrometer was 600g/mm, and the grating of Raman spectrometer was 1800g/mm. X-ray diffraction (XRD) of perovskite films and hybrid films were measured by a Bruker D8 ADVANCE diffractometer with Cu $K\alpha$ (λ = 1.5406 Å) radiation. Top-view SEM (scanning electron microscopy) images were obtained by FEI Nova Nano 630, FESEM. The absorption spectra of pristine WS_2 film, perovskite film and hybrid systems were obtained by an Agilent Cary 5000 UV-Visible-NIR spectrometer. The photodetector current voltage (IV) curves were measured by a Keithley 4200 Semiconductor Parametric Analyzer and Sigmotone Micoro manipulator S-1160 probe station. The photodetector on-off curve was obtained by Keithley 2400 Parametric Analyzer, and the transient time curve was recorded by a digital oscilloscope (Tektronix TDS2024C). Light source was provided and calibrated by LED Drive from Metrohm Autolab and optical bench, white light with different intensity was provided by LEDNW LED cover. Noise current of photodetector was measured directly by the Lock-in amplifier (SR830) and Temperature-dependent PL measurements from 77 K to 300 K were carried out by Horiba Aramis Raman microscope.

References

1 Hu, X. *et al.* High-Performance Flexible Broadband Photodetector Based on Organolead Halide Perovskite. *Adv. Funct. Mater.* **24**, 7373-7380 (2014).

2 Burschka, J. *et al.* Sequential deposition as a route to high-performance perovskite-sensitized solar cells. *Nature* **499**, 316-319 (2013).

3 Marchioro, A. *et al.* Unravelling the mechanism of photoinduced charge transfer processes in lead iodide perovskite solar cells. *Nat. Photonics* **8**, 250-255 (2014).

4 Sun, S. Y. *et al.* The origin of high efficiency in low-temperature solution-processable bilayer organometal halide hybrid solar cells. *Energ Environ Sci* **7**, 399-407 (2014).

5 Stranks, S. D. *et al.* Electron-Hole Diffusion Lengths Exceeding 1 Micrometer in an Organometal Trihalide Perovskite Absorber. *Science* **342**, 341-344 (2013).

6 Xing, G. C. *et al.* Long-Range Balanced Electron- and Hole-Transport Lengths in Organic-Inorganic CH3NH3PbI3. *Science* **342**, 344-347 (2013).

7 Shi, D. *et al.* Low trap-state density and long carrier diffusion in organolead trihalide perovskite single crystals. *Science* **347**, 519-522 (2015).

8 Liu, M. Z., Johnston, M. B. & Snaith, H. J. Efficient planar heterojunction perovskite solar cells by vapour deposition. *Nature* **501**, 395-398 (2013).

9 Ball, J. M., Lee, M. M., Hey, A. & Snaith, H. J. Low-temperature processed meso-superstructured to thin-film perovskite solar cells. *Energ Environ Sci* **6**, 1739-1743 (2013).

10 Lee, M. M., Teuscher, J., Miyasaka, T., Murakami, T. N. & Snaith, H. J. Efficient Hybrid Solar Cells Based on Meso-Superstructured Organometal Halide Perovskites. *Science* **338**, 643-647 (2012).

11 Tan, Z. K. *et al.* Bright light-emitting diodes based on organometal halide perovskite. *Nat. Nanotechnol.* **9**, 687-692 (2014).

12 Ning, Z. J. *et al.* Quantum-dot-in-perovskite solids. *Nature* **523**, 324-328 (2015).

13 Li, G. R. *et al.* Efficient Light-Emitting Diodes Based on Nanocrystalline Perovskite in a Dielectric Polymer Matrix. *Nano Lett.* **15**, 2640-2644 (2015).

14 Priante, D. *et al.* The recombination mechanisms leading to amplified spontaneous emission at the true-green wavelength in CH3NH3PbBr3 perovskites. *Appl. Phys. Lett.* **106**, 081902 (2015).

15 Deschler, F. *et al.* High Photoluminescence Efficiency and Optically Pumped Lasing in Solution-Processed Mixed Halide Perovskite Semiconductors. *J Phys Chem Lett* **5**, 1421-1426 (2014).

16 Xing, G. C. *et al.* Low-temperature solution-processed wavelength-tunable perovskites for lasing. *Nat. Mater.* **13**, 476-480 (2014).

17 Dou, L. T. *et al.* Solution-processed hybrid perovskite photodetectors with high detectivity. *Nat. Commun.* **5**, 5404 (2014).

18 Lee, Y. *et al.* High-Performance Perovskite-Graphene Hybrid Photodetector. *Adv. Mater.* **27**, 41-46 (2015).

19 Guo, Y. L., Liu, C., Tanaka, H. & Nakamura, E. Air-Stable and Solution-Processable Perovskite Photodetectors for Solar-Blind UV and Visible Light. *J Phys Chem Lett* **6**, 535-539 (2015).

20 Dong, R. *et al.* High-Gain and Low-Driving-Voltage Photodetectors Based on Organolead Triiodide Perovskites. *Adv. Mater.* **27**, 1912-1918 (2015).

21 Fang, Y. J. & Huang, J. S. Resolving Weak Light of Sub-picowatt per Square Centimeter by Hybrid Perovskite Photodetectors Enabled by Noise Reduction. *Adv. Mater.* **27**, 2804-2810 (2015).

22 Horvath, E. *et al.* Nanowires of Methylammonium Lead Iodide Prepared by Low Temperature Solution-Mediated Crystallization. *Nano Lett.* **14**, 6761-6766 (2014).

23 Chen, Q. *et al.* Planar Heterojunction Perovskite Solar Cells via Vapor-Assisted Solution Process. *J. Am. Chem. Soc.* **136**, 622-625 (2014).

24 Xiao, Z. G. *et al.* Efficient, high yield perovskite photovoltaic devices grown by interdiffusion of solution-processed precursor stacking layers. *Energ Environ Sci* **7**, 2619-2623 (2014).

25 Jeon, N. J. *et al.* Solvent engineering for high-performance inorganic-organic hybrid perovskite solar cells. *Nat. Mater.* **13**, 897-903 (2014).

26 Chen, C. W. *et al.* Efficient and Uniform Planar-Type Perovskite Solar Cells by Simple Sequential Vacuum Deposition. *Adv. Mater.* **26**, 6647-6652 (2014).

27 Liu, D. Y. & Kelly, T. L. Perovskite solar cells with a planar heterojunction structure prepared using room-temperature solution processing techniques. *Nat. Photonics* **8**, 133-138 (2014).

28 Konstantatos, G. & Sargent, E. H. Nanostructured materials for photon detection. *Nat. Nanotechnol.* **5**, 391-400 (2010).

29 McDonald, S. A. *et al.* Solution-processed PbS quantum dot infrared photodetectors and photovoltaics. *Nat. Mater.* **4**, 138-142 (2005).

30 Kung, P. *et al.* Kinetics of photoconductivity in n-type GaN photodetector. *Appl. Phys. Lett.* **67**, 3792-3794 (1995).

31 Xia, F. N., Mueller, T., Lin, Y. M., Valdes-Garcia, A. & Avouris, P. Ultrafast graphene photodetector. *Nat. Nanotechnol.* **4**, 839-843 (2009).

32 Chang, Y. H. *et al.* Monolayer MoSe2 Grown by Chemical Vapor Deposition for Fast Photodetection. *ACS Nano* **8**, 8582-8590 (2014).

33 Li, F. *et al.* Ambipolar solution-processed hybrid perovskite phototransistors. *Nat. Commun.* **6**, 8238 (2015).

34 Chhowalla, M. *et al.* The chemistry of two-dimensional layered transition metal dichalcogenide nanosheets. *Nat. Chem.* **5**, 263-275 (2013).

35 Wu, T. & Zhang, H. Piezoelectricity in Two-Dimensional Materials. *Angew Chem Int Edit* **54**, 4432-4434 (2015).

36 Zhang, Y. *et al.* Single-Layer Transition Metal Dichalcogenide Nanosheet-Based Nanosensors for Rapid, Sensitive, and Multiplexed Detection of DNA. *Adv. Mater.* **27**, 935-939 (2015).

37 Jo, S., Ubrig, N., Berger, H., Kuzmenko, A. B. & Morpurgo, A. F. Mono- and Bilayer WS2 Light-Emitting Transistors. *Nano Lett.* **14**, 2019-2025 (2014).

38 Zhang, Y. *et al.* Controlled Growth of High-Quality Monolayer WS2 Layers on Sapphire and Imaging Its Grain Boundary. *ACS Nano* **7**, 8963-8971 (2013).
39 Zhang, W. J. *et al.* Ultrahigh-Gain Photodetectors Based on Atomically Thin Graphene-MoS2 Heterostructures. *Sci Rep-Uk* **4**, 3826 (2014).
40 Berkdemir, A. *et al.* Identification of individual and few layers of WS2 using Raman Spectroscopy. *Sci Rep-Uk* **3**, 1755 (2013).
41 Zhu, B. R., Chen, X. & Cui, X. D. Exciton Binding Energy of Monolayer WS2. *Sci Rep-Uk* **5**, 9218 (2015).
42 Stoumpos, C. C., Malliakas, C. D. & Kanatzidis, M. G. Semiconducting Tin and Lead Iodide Perovskites with Organic Cations: Phase Transitions, High Mobilities, and Near-Infrared Photoluminescent Properties. *Inorg. Chem.* **52**, 9019-9038 (2013).
43 Huo, N. J. *et al.* Photoresponsive and Gas Sensing Field-Effect Transistors based on Multilayer WS2 Nanoflakes. *Sci Rep-Uk* **4**, 5209 (2014).
44 Koppens, F. H. L. *et al.* Photodetectors based on graphene, other two-dimensional materials and hybrid systems. *Nat. Nanotechnol.* **9**, 780-793 (2014).
45 Lopez-Sanchez, O., Lembke, D., Kayci, M., Radenovic, A. & Kis, A. Ultrasensitive photodetectors based on monolayer MoS2. *Nat. Nanotechnol.* **8**, 497-501 (2013).
46 Radisavljevic, B., Radenovic, A., Brivio, J., Giacometti, V. & Kis, A. Single-layer MoS2 transistors. *Nat. Nanotechnol.* **6**, 147-150 (2011).
47 D'Innocenzo, V. *et al.* Excitons versus free charges in organo-lead tri-halide perovskites. *Nat. Commun.* **5**, 3586 (2014).
48 Wu, K. W. *et al.* Temperature-dependent excitonic photoluminescence of hybrid organometal halide perovskite films. *PCCP* **16**, 22476-22481 (2014).
49 Even, J., Pedesseau, L. & Katan, C. Analysis of Multivalley and Multibandgap Absorption and Enhancement of Free Carriers Related to Exciton Screening in Hybrid Perovskites. *J. Phys. Chem. C* **118**, 11566-11572 (2014).
50 Savenije, T. J. *et al.* Thermally Activated Exciton Dissociation and Recombination Control the Carrier Dynamics in Organometal Halide Perovskite. *J Phys Chem Lett* **5**, 2189-2194 (2014).
51 Gutierrez, H. R. *et al.* Extraordinary Room-Temperature Photoluminescence in Triangular WS2 Monolayers. *Nano Lett.* **13**, 3447-3454 (2013).

Chapter 6

One-Dimensional SWCNT Hybrid with Perovskite with High Mobility and High Photoresponse

Synopsis

Benefiting from their extraordinary physical properties, methylammonium lead halide perovskites (PVKs) have attracted significant attention in optoelectronics. However, the PVK-based devices suffer from low carrier mobility and high operation voltage. Here, we utilize sorted semiconducting single-walled carbon nanotubes (95% s-SWCNT) to enhance the performance of thin-film transistors (TFTs) based on the mixed-cation perovskite $(MA_{1-x}FA_x)Pb(I_{1-x}Br_x)_3$, enabling mixed-dimensional solution-processed electronics with high mobility (32.25 cm^2/Vs) and low voltage (3 V) operation. The resulting mixed-dimensional PVK/SWCNT TFTs possess ON/OFF ratios on the order of 10^7, enabling the fabrication of high-gain inverters.

6.1 Introduction

In addition to their well-known application in photovoltaic devices (certified power conversion efficiencies exceeding 23% [1]), PVKs have been explored for other devices such as photodetectors [2-5], light-emitting devices[6], and field-effect transistors [7-10]. In particular, light-emitting diodes based on organometallic halide PVKs have shown a high external quantum efficiency of 20.7% with a power conversion efficiency of 12% [6]. PVK-based photodetectors have also exhibited excellent performance including high sensitivities and wide spectral responses ranging from ultraviolet-visible-near-infrared (UV-NIR) frequencies [3] to X rays [11-13] and gamma rays [14-15]. In PVKs, strong inter-band transitions [16] are accompanied by dominant bimolecular recombination below the Langevin limit [17-18] and low trap densities [19-20]. These characteristics, combined with the photo-recycling effect [21-22], give rise to high light absorption and long carrier diffusion lengths [23], suggesting a high potential for solution-processed PVK optoelectronics.

Table 6.1 A comparison of different studies of transistors and inverters based on PVKs.

Materials	Dielectric	Mobility (cm^2/Vs)	Threshold voltage (V)	ON/OFF ratio	inverter Gain	Reference (year)
$(PEA)_2SnI_4$	SiO_2	0.29 – 1.7	~ -5	~ 10^6		
$MAPbI_{3-x}Cl_x$	SiO_2	1.0 – 1.3	5 – 10	~ 10^5		Li et al. (2015)
$MAPbI_3$/PEIE	Cytop/ SiO_2	3 – 5	~ -3	~ 10^5		Friend *et al.* (2017)
$MAPbI_3$ nanoplate	SiO_2	1 – 2.5	~ 15	Nearly 10^6		Duan *et al.* (2015)
$MAPbI_3/WSe_2$	SiO_2		~ 20	10^6		Duan *et al.* (2015)
$Cs_x(MA_{0.17}FA_{0.83})_{1-x}Pb(Br_{0.17}I_{0.83})_3$	SiO_2	2.02 – 2.39	~ 50	10^4 – 10^6	23 (V_{DD}=80 V)	(2016)
$MAPbI_{3-x}Cl_x$/unsorted CNTs	SiO_2	108.7–595.3	~ 2	10^2		Li *et al.* (2017)

PVKs are also expected to accelerate the development of solution-processed electronics. **Table 6.1** provides a list of representative studies on PVK-based thin-film transistors (TFTs). In one of the first studies, Mitzi *et al.* reported a $(PEA)_2SnI_4$ TFT with a mobility of 0.6 cm^2/Vs [24]. More recent work has focused on $MAPbI_{3-x}Cl_x$ TFTs with reported mobilities in the range of 1.01-1.24 cm^2/Vs [10]. Several efforts have attempted to improve TFT performance by optimizing PVK composition [25], fabricating hybrid bilayers [26], and using single-crystal microplates [9]. A particularly powerful approach is to interface PVK with high-mobility nanomaterials such as graphene [27-28] and carbon nanotubes (CNTs) [7]. However, these devices have thus far suffered from low ON/OFF ratios and high threshold voltages [29-30], limiting their electronic applications.

Single-walled carbon nanotubes (SWCNTs) are favorable for solution-processed TFT applications because of their electronic tunability [31], high carrier mobility[32], exceptional thermal conductivity and mechanical properties [33], and inherent dispersibility in organic solvents compatible with PVK precursor solutions [34]. With high mechanical resilience, SWCNTs have also sparked great interest for stretchable and flexible electronics [35]. Most SWCNT synthetic methods result in a 1:2 ratio of metallic (m-SWCNTs) to semiconducting (s-SWCNTs) species [36-37]. Unlike m-SWCNTs that have no bandgap [38], s-SWCNTs have a tunable bandgap and can form type-II heterostructures with PVKs [39-40]. Previously, our group reported heterostructures based on PVKs and unsorted SWCNTs [39-40], but these devices suffered from high OFF currents (10 mA at 5 V) and low ON/OFF ratios of 10^2, which is a result of residual m-SWCNTs in the mixture [41]. Isolation and enrichment of s-SWCNTs can be accomplished by several methods such as gel-permeation chromatography [42], density gradient ultracentrifugation (DGU) [43], and selective dispersion with conjugated polymers [44]. However, high-purity s-SWCNTs have not yet been utilized in PVK TFTs, suggesting an opportunity for improved performance.

Here, the DGU technique is used to enrich polychiral s-SWCNTs produced by the high pressure carbon monoxide (HiPCo) growth method, allowing the production of TFTs using mixed-cation PVKs $(MA_{1-x}FA_x)Pb(I_{1-x}Br_x)_3$ and sorted s-SWCNTs. Compared to previous devices based on unsorted SWCNTs, the OFF currents of TFTs with sorted s-SWCNTs are significantly suppressed by 4 orders of magnitude, resulting in record-high ON/OFF ratios (10^7). Furthermore, we employed a high-κ gate dielectric layer (HfO_2) to dramatically reduce the TFT threshold voltage from ~ 60 V (SiO_2 devices) to ~3 V, enabling low-voltage operation. A thin self-assembled monolayer (SAM) of ethoxylated polyethyleneimine (PEIE) was further used to enhance device stability. To demonstrate the utility of these TFTs for digital electronics, we fabricated a solution-processed inverter with desirable characteristics based on PVKs and s-SWCNTs. As a mixed-dimensional heterostructure[45], the PVK/s-SWCNT composite exhibits unique physical phenomena such as a transition from high-temperature ambipolar to low-temperature unipolar charge transport behavior, which is a result of competing carrier trapping and emission.

6.2 Result and Discussion

Both PVKs and SWCNTs were prepared in solution. The inset in **Figure 6.1** presents a photograph of the PVK precursor solutions with different concentrations of SWCNTs (pristine, 0.1 wt.%, 0.3 wt.%, and 0.5 wt.%). As our previous work demonstrated [7], the PVK precursor solution of polar, non-hydrogen-bonding Lewis base solvents, N,N-dimethylformamide (DMF) and dimethyl sulfoxide (DMSO), allows for the stable, homogeneous dispersion of SWCNTs [39]. We used a two-step, anti-solvent engineering method to deposit the PVK films [46-47], which is illustrated in **Figure 6.1a**. Different from our previous work [7], here we used precursors of MAI, FABr, PbI_2, and $PbBr_2$ mixed in DMF/DMSO (4:1) to prepare mixed-cation and mixed-halide $(MA_{1-x}FA_x)Pb(I_{1-x}Br_x)_3$ perovskites. Thin films based on mixed-cation and mixed-halide PVKs were found to be more stable and possess higher carrier mobility than the commonly used $MAPbI_3$ perovskite [48-49]. The improved properties of mixed-cation PVKs could be a result of their cubic structure with a tolerance factor close to 1. After spin-coating and annealing at 100 °C for 1 hour, dense and uniform polycrystalline PVK thin films embedded with SWCNTs were achieved. More experimental details are given in the experimental section.

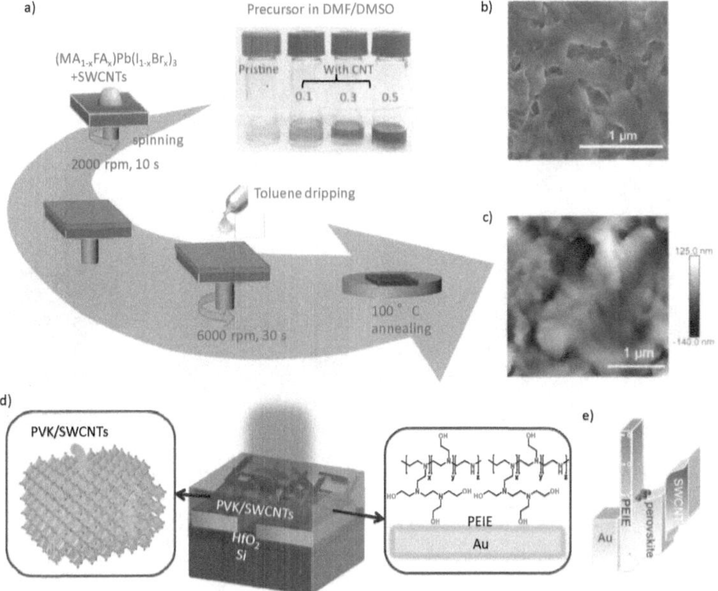

Figure 6.1 Schematic of fabrication and characterization of PVK/SWCNT transistors. a) Photograph of the PVK precursor solutions with different concentrations of SWCNTs and schematic illustration of the two-step solvent engineering process for the fabrication of films composed of mixed-cation PVK and SWCNTs. b) Top-view SEM image of a typical PVK/s-SWCNT film with a SWCNT concentration of 0.1 mg/mL. c) AFM image of the as-prepared PVK/SWCNT film. d) Device structure of the PVK/SWCNT phototransistor, accompanied by schematic representations of the PVK/SWCNT channel and the molecular PEIE self-assembled layer on the top of Au electrodes. The channel length (L) is 50 µm, and the channel width (W) is 1 mm. e) Band alignment of the mixed-dimensional PVK/s-SWCNT channel.

Figures 6.1b, c show the top-view scanning electron microscopy (SEM) and atomic force microscopy (AFM) images, respectively, of the as-prepared PVK/SWCNT thin film. The SWCNTs appear to be uniformly distributed in the PVK thin film with no evidence of agglomerates or bundles. The solvent molecules and ionic species in the PVK precursor solutions likely interact with the SWCNTs through van der Waals and Coulombic/steric forces that act to reduce aggregation and promote uniform dispersion [50]. Furthermore, the PVK/SWCNT thin film showed continuous surface coverage without pinholes.

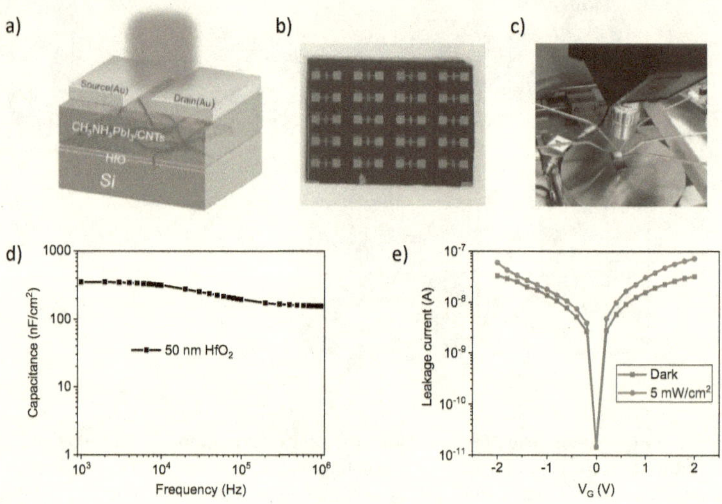

Figure 6.2 (a) Device structure and (b) photograph of the PVK/SWCNT TFTs. (c) Set-up of the inverter measurement. (d) Frequency-dependent capacitance of the HfO_2 dielectric layer. (e) Leakage current of the HfO_2 layer.

The bottom-gate, bottom-contact PVK/SWCNT TFT is illustrated in **Figure 6.1d**. HfO_2 with a thickness of 50 nm deposited by atomic layer deposition (ALD) is used as the gate dielectric layer, and the heavily n-doped Si substrate is used as the back gate [51]. The relative permittivity (ε) of the HfO_2 dielectric was characterized to be 19 (**Figure 6.2**), which is substantially higher than that of SiO_2 (ε=3.7). Au electrodes with a thickness of 80 nm were sputtered on the HfO_2 layer. The channel length between electrodes was 50 μm, and the channel width was 1 mm. The thin HfO_2 layer with high permittivity affords the low-bias operation of the hybrid PVK/SWCNT electronic devices. As illustrated in **Figure 6.1d**, after the electrode deposition, a thin (approximately 5 nm) self-assembled monolayer (SAM) of ethoxylated polyethyleneimine (PEIE) was spin-coated on the substrate. This layer provides several benefits in terms of device stability and performance. First, it protects the Au electrodes from reacting directly with the PVK layer.

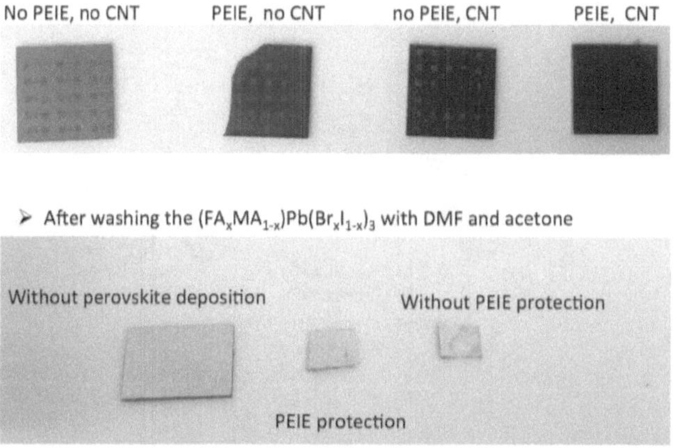

Figure 6.3 Upper images are the transistors with and without PEIE protection, and with and without SWCNTs. Lower images are Au films (80 nm) after deposition of PVKs and washing with DMF and acetone.

As demonstrated in **Figure 6.3**, without the protective PEIE layer, the Au electrode was partially etched. X-ray photoelectron spectroscopy (XPS) taken on the Au surface without the PEIE protection (**Figure 6.4**) show a blue-shift of the Au $4f_{7/2}$ and $4f_{5/2}$ peaks [52], providing additional evidence of reaction between the Au surface and the perovskite layer. Second, the self-assembled dipoles in the thin PEIE layer likely improve the quality of the PVK layer by passivation of the interface [53]. Third, the PEIE layer facilitates the PVK growth and enhances the substrate adhesion of the PVK grains (**Figure 6.3**). Finally, from an energetic point of view, PEIE increases the Fermi level of Au by 0.1 eV, thereby facilitating charge injection by lowering the height of the Schottky barrier between the Au electrode and the semiconductor thin film [8].

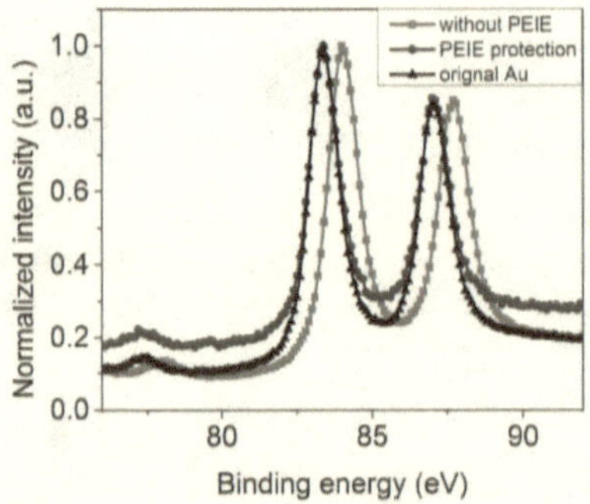

Figure 6.4 XPS spectra of Au electrodes after the deposition of PVKs and washing with DMF and acetone.

The mixed-dimensional PVK/SWCNT heterostructure is illustrated in the inset of **Figure 6.1d**. XPS spectra in **Figure 6.5** confirm the existence of SWCNTs in the PVK film. **Figure 6.1e** schematically shows the electronic band alignment between the

SWCNTs and PVK in the channel layer, together with the Au electrodes and PEIE interlayer. Because of the type-II band alignment, effective carrier separation is expected at the PVK/SWCNT interface, enhancing charge transport and mobility by reducing Coulomb scattering and recombination [39].

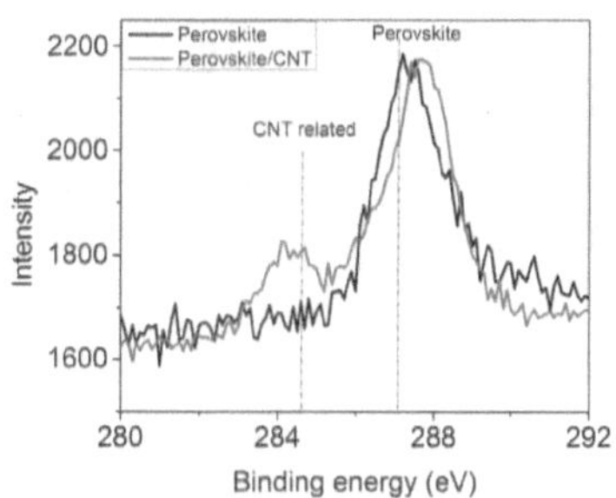

Figure 6.5 XPS spectra of pristine PVK and PVK/SWCNT films in the energy range of 280 eV to 292 eV.

We first characterized the SWCNT samples using optical absorption. **Figure 6.6a** shows the optical absorbance spectra of the metallic-enriched, semiconducting-enriched, and unsorted SWCNTs. Absorbance of both sorted semiconducting and unsorted dispersions is suppressed in the wavelength range from 440 to 600 nm, the region associated with the first-order excitations (M11) of the SWCNT metallic species. Further analysis revealed that the sorted s-SWCNTs consist of approximately 95% semiconducting species, while the unsorted s-SWCNTs consist of approximately 80% semiconducting species. Overall, these results are consistent with previous reports on semiconducting enrichment of HiPCo SWCNTs by DGU [43].

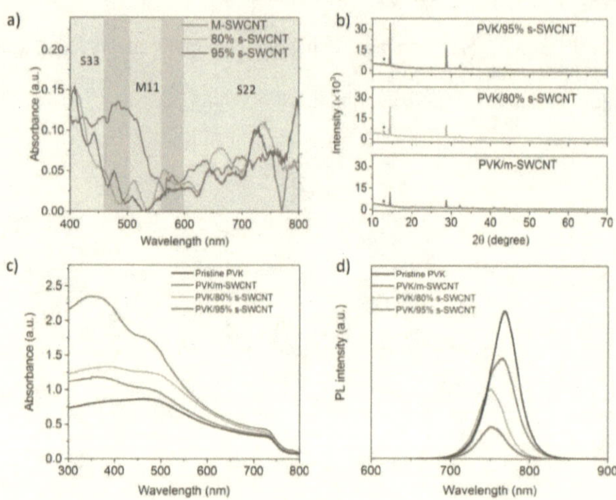

Figure 6.6 Optical and structural characterizations. a) UV-Vis spectra of metallic-enriched, semiconducting-enriched, and unsorted SWCNTs. M11 is the peak associated with the SWCNT metallic species, and S22/S33 represent the SWCNT semiconducting species. b) XRD patterns of the PVK/SWCNT films with different types of SWCNTs (95% semiconducting, 80% semiconducting, and metallic SWCNTs). c) UV-Vis absorption spectra of the PVK/SWCNT films with different types of SWCNTs. d) Corresponding photoluminescence spectra of different PVK/SWCNT films.

To further investigate the structure of the PVK/SWCNT film, X-ray diffraction (XRD) was performed. **Figure 6.6b** shows the XRD patterns of the PVK films with different types of SWCNTs with 0.02 w.t.% concentration. Peaks at 14.21°, 28.51°, 43.82° can be assigned to the (110), (220), (330) crystallographic planes of $(MA_{1-x}FA_x)Pb(I_{1-x}Br_x)_3$, and the peak at 13.94° belongs to residual PbI_2, which may help passivate the PVK grains [54]. Interestingly, under the same measurement conditions, the X-ray diffraction peaks of the sorted PVK/s-SWCNT films were sharper and stronger than those of PVK/m-SWCNT films. A zoom-in comparison of the (110) XRD peaks of three samples is presented in **Figure 6.7**. Our result indicates that the uniformity of sorted s-SWCNTs enhances the crystallinity of the PVK films, which is consistent with the larger grain size observed in the SEM images (**Figure 6.8**). Because most structural defects exist at the grain boundaries [55], we expect reduced charge scattering and improved transport properties in the PVK films with sorted s-SWCNTs.

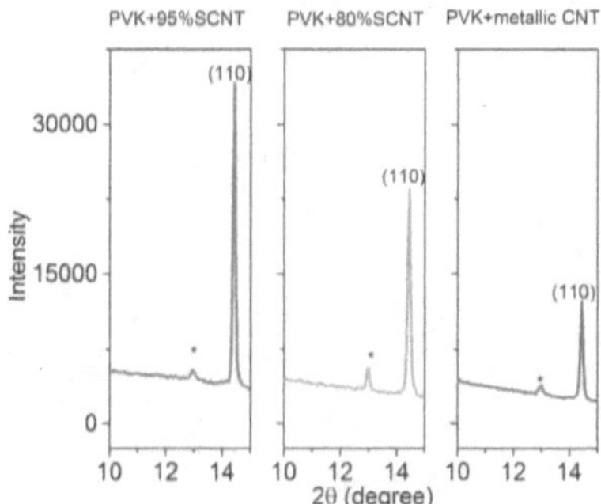

Figure 6.7 Enlarged XRD patterns (10° to 15°) of PVK with semiconducting-enriched SWCNTs, unsorted SWCNTs, and metallic-enriched SWCNTs with the concentrations of the different SWCNTs being held constant at 0.1 mg/mL. PVK films with semiconducting-enriched SWCNTs shows better crystallinity with lower impurity peak (*) and higher (110) peak.

The structural modification of the PVK films by the s-SWCNTs also impacts the optical properties of the hybrid films. **Figure 6.6c** shows the UV-vis absorption spectra of

PVK films with different types of SWCNTs. All films exhibit an absorption edge at 770 nm, which is consistent the PVK bandgap [25, 48]. The optical absorption in the PVK/SWCNT films is enhanced in the short wavelength range (300-500 nm) compared to the pristine PVK film. This enhancement in absorption is expected because the addition of SWCNTs improves the morphology and crystallinity of the PVK film [39].

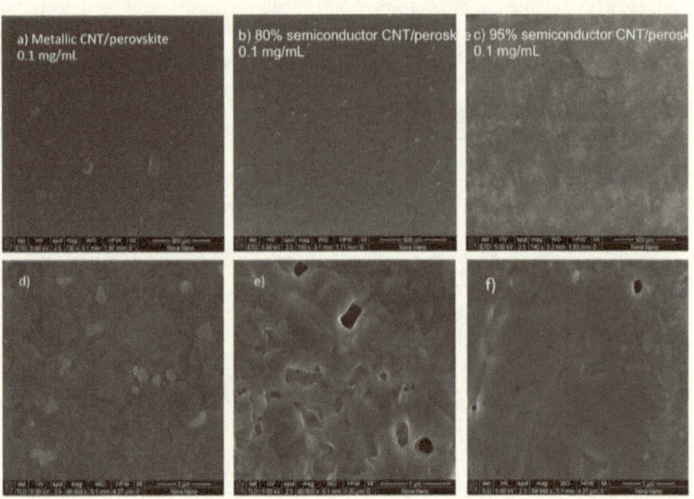

Figure 6.8 SEM images of PVK films with (a) metallic-enriched SWCNTs, (b) unsorted SWCNTs, and semiconducting-enriched SWCNTs with the same concentration of 0.1 mg/mL. (d,e,f) are the enlarged SEM images of the same set of samples

As shown in **Figure 6.6d**, substantial photoluminescence (PL) quenching was observed in the hybrid PVK/SWCNT film, confirming the formation of a type-II heterojunction with efficient exciton dissociation at the interfaces. The most significant PL quenching occurs in the hybrid PVK film embedded with enriched s-SWCNTs. Besides the PL quenching, a blue shift (25 nm or 53 meV) of the PL peak is observed with increasing purity of s-SWCNTs, implying the suppression of in-gap trap states [19].

We fabricated bottom-contact, bottom-gate PVK/SWCNT TFTs with Au electrodes and a HfO_2 gate dielectric layer. After testing dozens of devices, the thickness of the PVK film was optimized to be 140 nm (**Figure 6.9**). If the channel film was too thin, it was discontinuous with poor morphology. On the other hand, if the film was too thick, the gate bias effect was reduced due to screening.

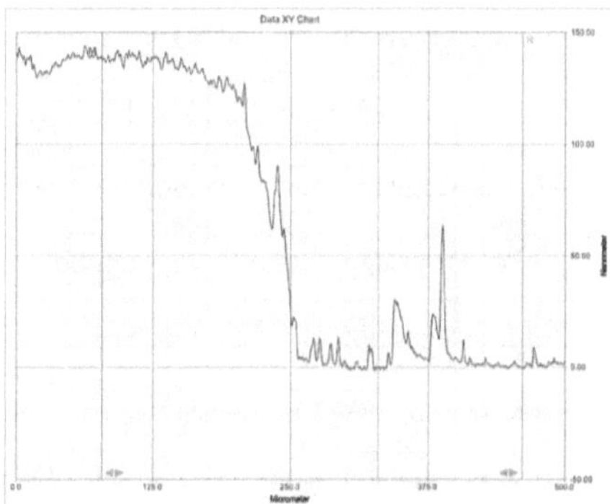

Figure 6.9 Profilometry data show that the thickness of the optimized channel is ~140 nm.

Figure 6.10a is the transfer curve (I_{DS} vs. V_g) of the TFT with a pristine PVK channel. We swept the back-gate voltage V_g at a rate of 0.5 V/s, while maintaining the source-drain voltage V_{DS} at a fixed bias (-1 V). The pristine PVK TFT shows a clear ambipolar behavior with both electron and hole transport, which is consistent with previous reports [7, 56]. However, the source-drain current (I_{DS}) is lower than 0.1 μA, implying overall poor charge transport for pristine PVK. The hole mobility (μ) was extracted using the following relationship [7]:

$$I_{DS} = \mu C_i \frac{W}{L}\left((V_g - V_{TH})V_{DS} - \frac{V_{DS}^2}{2}\right) \tag{6.1}$$

where C_i is gate capacitance (355.6 nF/cm^2). The hole mobility of the pristine PVK film was calculated to be 0.033 cm^2/V.

The transport characteristics of the PVK/SWCNT TFTs sensitively depend on the SWCNT concentration. **Figures 6.10b** and **6.10c** show the transfer curves of PVK films with s-SWCNTs of 0.1 mg/mL (0.02 wt.%, PVK+SWCNT-01) and 0.5 mg/mL (0.1 wt.%, PVK+SWCNT-05), respectively. The source-drain current increases dramatically with increasing SWCNT concentration, achieving a high hole mobility of 18.3 cm^2/Vs at an unsorted s-SWCNT concentration of 0.02 wt.%. The transfer curve of PVK+SWCNT-01 remains ambipolar with suppressed hysteresis, which implies suppressed trap states and improved sample quality after the addition of high-purity SWCNTs. The high-mobility SWCNTs embedded in the PVK matrix provide pathways for hole transport, and the efficient charge separation reduces electron-hole recombination at grain boundaries. In contrast, the PVK+SWCNT-05 device showed a p-type transfer curve with a high OFF current. Although the hole mobility is still high (8.9 cm^2/Vs), it suffers from a relatively low ON/OFF ratio (less than one order of magnitude). At this high concentration, the

SWCNT percolation threshold is exceeded, resulting in transport being dominated by p-type SWCNTs in addition to increased screening that suppresses the ON/OFF ratio.

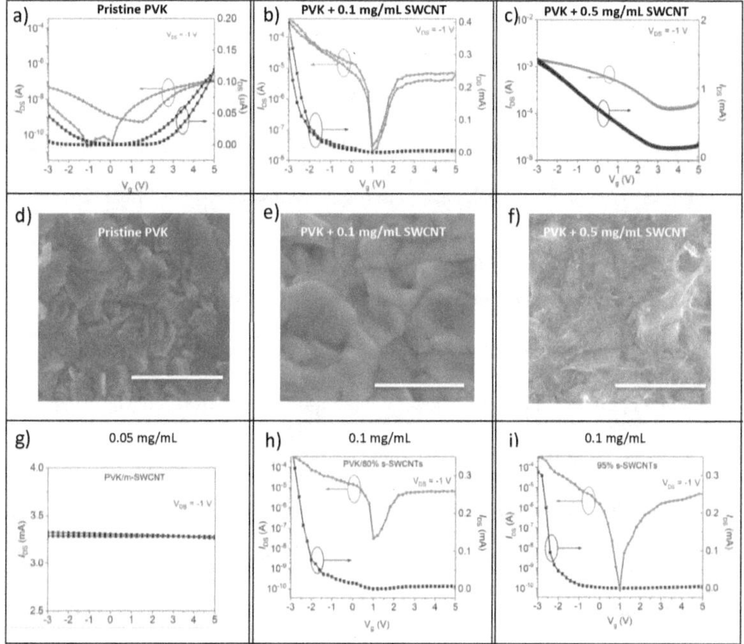

Figure 6.10 Performance of PVK/SWCNT TFTs. Transfer characteristics of (a) the pristine mixed-cation PVK TFT and the devices with (b) 0.1 mg/mL 80% s-SWCNTs and (c) 0.5 mg/mL 80% s-SWCNTs. (d-f) Corresponding top-view SEM images of the PVK films with different concentrations of SWCNTs. The scale bars are 1 μm. (g) Transfer curve of the PVK/m-SWCNTs transistor with a metallic-enriched SWCNT concentration of 0.05 mg/mL. (h,i) transfer curves of the PVK/SWCNT TFTs with 0.1 mg/mL of unsorted and semiconducting-enriched (95%) SWCNTs, respectively.

Figure 6.10d,e,f show the top-view SEM images of the PVK films with different concentrations of SWCNTs. Faceted grains were observed in the pristine film (**Figure 6.10d**), and they become larger when SWCNTs with a concentration of 0.1 mg/mL were added to the samples (**Figure 6.10e**). The high crystallinity of the PVK/SWCNT film benefit both the optical absorption (**Figure 6.6c**) and the charge transport (**Figure 6.10b**). On the other hand, as shown in **Figure 6.10f**, at the highest SWCNT concentration (0.5 mg/mL), large SWCNT bundles are observed, which explains the high OFF current in the TFT. In order to determine the effect of SWCNT type, we compared the performance of TFTs with metallic-enriched, semiconducting-enriched, and unsorted SWCNTs. **Figure 6.10g** shows the transfer curve of the PVK/m-SWCNT device, which is highly conductive with negligible gate tuning. In contrast, improved performance was observed in TFTs with unsorted (**Figure 6.10h**) and semiconducting-enriched (**Figure 6.10i**) SWCNTs. In particular, high-purity s-SWCNTs maintain the high ON current while suppressing the OFF current to a low level (10^{-10} A). The device with high-purity s-SWCNTs exhibits a hole mobility of 32.2 cm^2/Vs in linear region. Furthermore, an ON/OFF ratio of 10^7 was observed.

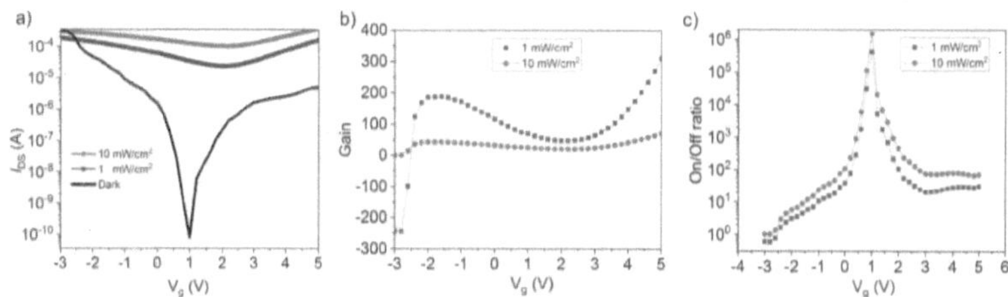

Figure 6.11 Photoresponsivity of the PVK/SWCNT TFT operated as a phototransistor. a) Transfer curves of the PVK/SWCNT phototransistor at different light irradiation conditions (dark, 1 mW/cm^2, and 10 mW/cm^2). b,c) Gain and ON/OFF ratio of the phototransistor.

It should be noted that all the transport measurements above were carried out in the dark and vacuum. Under light illumination, the off-current increases because of photo-induced carriers, leading to a decreased ON/OFF ratio. Nevertheless, as **Figure 6.11** shows, with the increase of OFF current, the PVK/SWCNT TFT can operate as a phototransistor with a gain of above 200 and an ON/OFF ratio of 10^6.

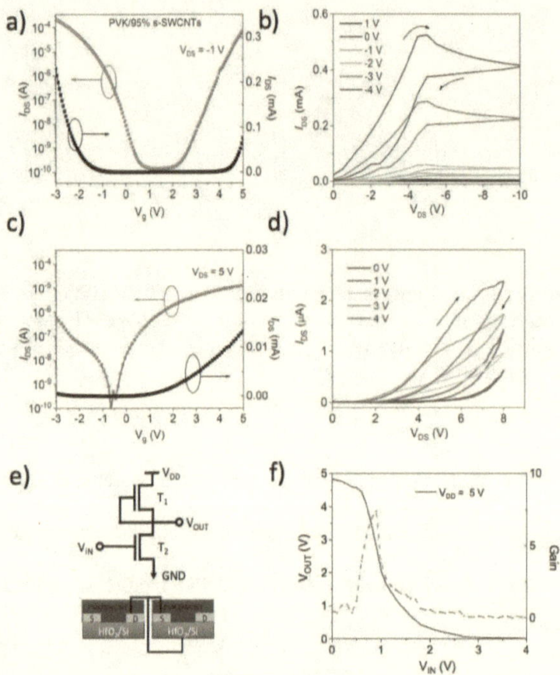

Figure 6.12 PVK/SWCNT inverter. a) Transfer curve and b) output curve of the p-type dominant PVK/SWCNT TFT with V_{DS} of -1 V. c) Transfer curve and d) output curve of the n-type dominant PVK/SWCNT TFT with V_{DS} of 5 V. e) Schematic illustration of the PVK/SWCNT inverter, consisting of two identical PVK/SWCNT TFTs (s-SWCNT concentration: 0.1 mg/mL). f) Input-output curve of the PVK/SWCNT inverters with drive voltages at 5 V. The dashed lines show the voltage-gain characteristics.

To illustrate the utility of the PVK/SWCNT TFTs in digital electronics, a PMOS inverter was fabricated and characterized as shown in **Figure 6.12**. Specifically, **Figures 6.12a** and **6.12b** show the transfer and the output curves, respectively, of the p-type dominant PVK/SWCNT TFT. The output curves exhibit saturation at ~ 5 V and a clockwise hysteresis, which may result from the trap states in the PVK film [8]. **Figures 6.12c** and **6.12d** show the transfer curve and the output curve of the n-type dominant PVK/SWCNT TFT with V_{DS} at 5 V. **Figure 6.12e** illustrates the schematic of the PVK/SWCNT inverter. The inverter was connected in a complementary configuration [25], consisting of two identical PVK/SWCNT TFTs. **Figure 6.12f** shows the voltage transfer characteristics (V_{OUT} vs V_{IN}) of the inverters with different supply voltages V_{DD}. At ~ 1 V, the input signal was inverted with sharp switching. As shown in **Figure 6.12f**, when the supply voltage is 5 V, the gain of the inverter can achieve the maximum value of 7.6. The trip voltage (~1 V) is close to the ideal trip voltage ($V_{DD}/2$), which leads to stable operation of the inverters. Compared to previous work [8], the use of HfO_2 dielectric layer lowers the threshold voltage below 3 V, making such PVK/SWCNT devices promising for low-voltage optoelectronics.

To gain insight on the carrier transport and trapping, we measured the TFT characteristics at variable temperatures (**Figures 6.13a** and **6.13b**). Four conductance regions of the PVK/SWCNT channel can be defined. As temperature increases, the transition from region (I) to (II) can be attributed to the phase transition (from tetragonal to orthorhombic) occurring at around 160 K in PVK films, which was reported to affect charge transport [57]. When the temperature further increases to room temperature, the conductance starts to decrease from region (II) to (III). One possible reason behind this transition is that ions defreeze at high temperature, and the ion screening and scattering effects limit the charge transport. Finally, region (IV) shows an emergence of electron conductance at high positive bias, and electrons that are trapped at low temperatures start to be released from ionized traps when the temperature increases above 160 K [58].

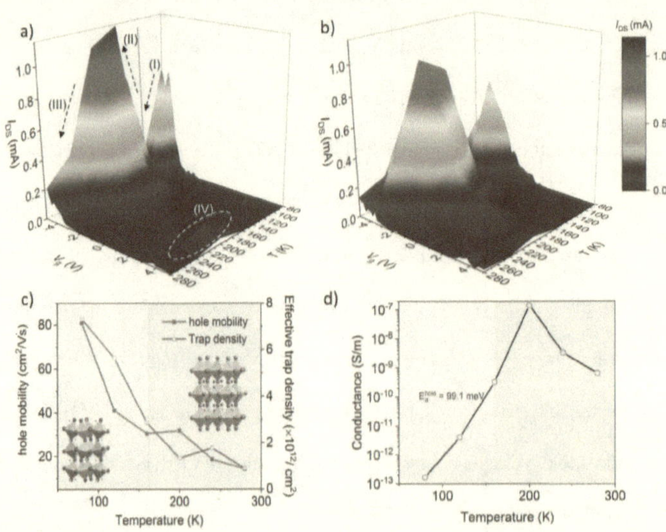

Figure 6.13 Temperature-dependent performance of the PVK/SWCNT TFT. Transfer curves of the PVK/SWCNT TFT measured at different temperatures with a) forward scanning and b) reverse scanning. c) Hole mobility and trap density extracted from the temperature-dependent transfer curves. Insets are the schematic of the crystal structures of perovskites at low temperature (orthorhombic) and room temperature (tetragonal). d) Conductance of the PVK/SWCNT channel at different temperatures.

As shown in **Figure 6.14a**, at a low temperature of 80 K, the transfer curve of the transistor shows an anti-clockwise hysteresis, and the p-type behavior becomes much pronounced. With temperature increasing to 120 K, the ON/OFF ratio slightly decreases while the p-type behavior is sustained (**Figure 6.14b**). At 160 K, the hysteresis loop significantly shrinks, indicating the decrease of effective trap density (**Figure 6.14c**). When the temperature continues increasing to 200 K, the shape of the hysteresis loop changes as a result of the emergence of n-type transport (**Figure 6.14d**). At such intermediate temperatures, electrons are involved in the carrier transport, although their role is still moderate compared to holes. This trend continues at 240 K (**Figure 6.14e**), and at 280 K the PVK/SWCNT channel shows ambipolar characteristics. As a general trend, with increasing temperature, the hysteresis loop continues shrinking, indicating the suppression of the effects of both electron and hole traps.

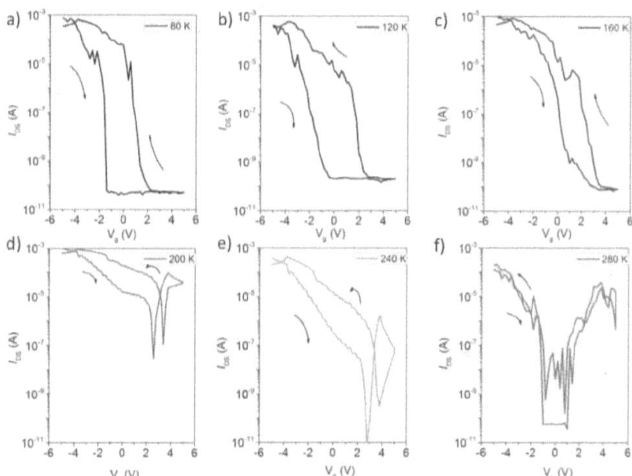

Figure 6.14 Temperature dependence of the PVK/SWCNT TFT. a-f) Transfer curves of the PVK/SWCNT TFT measured at different temperatures (80 K, 120 K, 160 K, 200 K, 240 K, and 280 K).

Since in the sub-threshold region channel transport can be tuned by the gate voltage without saturation, the PVK/SWCNT film is in the flat-band region and less affected by space charges [59]. In this region, the hole mobility can be extracted from

$$I_{DS} = \mu_{Linear} C_i \frac{W}{L}\left((V_g - V_{TH})V_{DS} - \frac{V_{DS}^2}{2}\right) \tag{6.3}$$

$$\mu_{Linear} = \frac{L}{C_i W V_{DS}} \frac{\partial I_{DS}}{\partial V_g}. \tag{6.4}$$

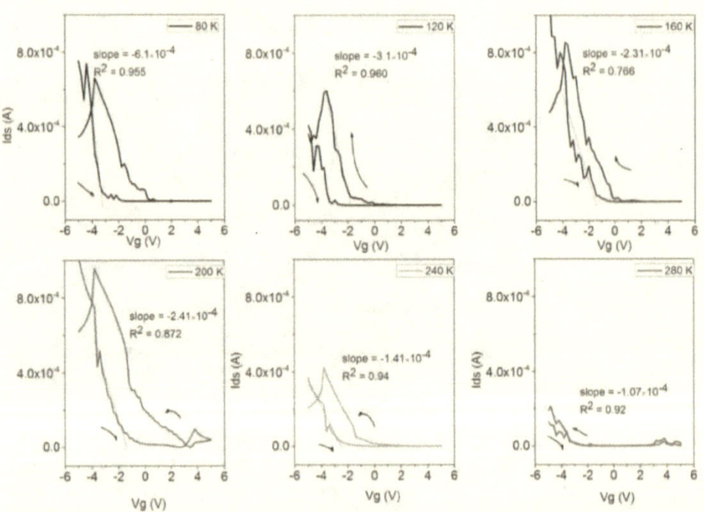

Figure 6.15 Transfer curves of PVK/SWCNT transistors at different temperatures.

The linear fittings are given in **Figure 6.15**. As shown in **Figure 6.13c**, the hole mobility decreases with increasing temperature, which is in line with the general behavior of semiconductors. The hole mobility of the transistor is calculated to be ~80 cm^2/Vs at 80 K. At higher temperatures, both phonon scattering and ion migration contribute to the scattering cross-section of holes. However, a plateau of hole mobility is observed at ~160 K, which coincides with the structural transition of perovskites as reported in previous studies [8, 57]. Furthermore, the total trap density can be determined by the hysteresis loop in the transfer curves, *i.e.*, $N_{TT} = C_{ox}\Delta V_{TH}/q$, in which $C_{ox}=\varepsilon_{ox}/t_{ox}$ is the oxide capacitance, q is the elementary charge, and ΔV_{TH} is the difference of V_{TH} when scanning in the forward and reverse directions [58]. As shown in **Figure 6.13c**, the effective trap density decreases with increasing temperature, which results from the higher kinetic energy of carriers and higher emission probability. The trap density is 7.3×10^{12} cm^{-2} at 80 K and decreases to 8.8 $\times10^{11}$ cm^{-3} at room temperature, which is quite close to the values reported for pure $MAPbI_3$ films [60], indicating that SWCNTs do not exacerbate defect trapping in PVK films.

Figure 6.13d shows the temperature-dependent conductivity (σ) of the PVK/SWCNT transistor channel at V_g = 2 V and V_{DS} = -1 V. The conductance was calculated by $\sigma = L/tWR$, in which L is the channel length (50 μm), W is the channel width (1000 μm), and t is the film thickness (140 nm). We observed a clear transition of conductance at 160 K. Since ion migration is suppressed at low temperatures and TFTs show p-type behavior when the temperature is below 160 K, we can assume holes as the dominant carriers. From fitting to the Arrhenius plot in **Figure 6.16**, we obtained the activation energy (E_a) of holes as 99.1 meV. First-principles calculation by Yin *et al.* predicted that the electronic activation energy of MA vacancies V_{MA} is as small as 50 meV [61]. Temperature-dependent transient response measurements performed by Li *et al.* obtained an activation energy of 120 ± 10 meV [62]. Four-probe Hall and SCLC measurements on perovskite single crystals reported by Adinolfi *et al.* revealed a hole trapping state peak at 0.1 eV and an electron trapping state peak at 0.2 eV [19].

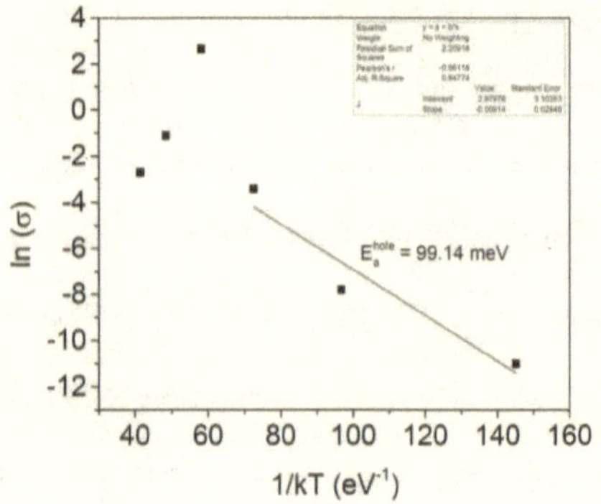

Figure 6.16 Plot of Ln (σ) vs 1/kT. From this plot, the activation energy is calculated to be 99.14 meV.

The experimentally obtained hole activation energy is approximately twice the theoretical prediction, which indicates additional charge trapping caused by disorder introduced in solution processing. Furthermore, the fact that the activation energy obtained here is comparable to previous reports on pure perovskite films implies that the incorporation of s-SWCNTs does not exacerbate the carrier trapping. Finally, **Figure 6.13d** shows that when the temperature is above 160 K, the conductance of films decreases with increasing temperature, which indicates enhanced charge scattering caused by ion migration at high temperatures.

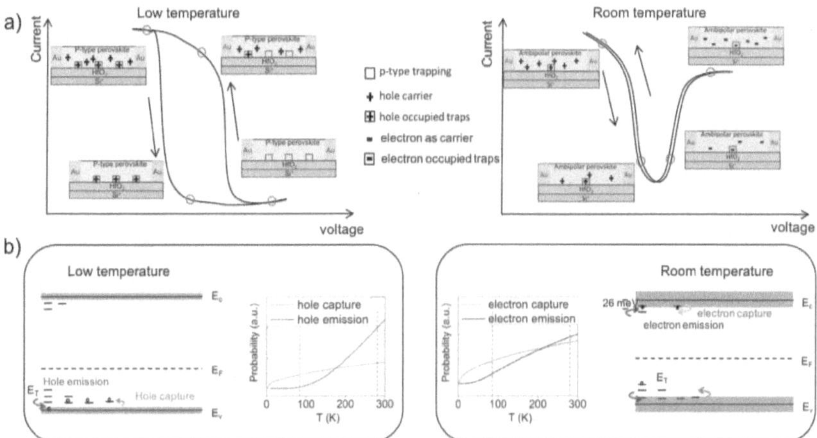

Figure 6.17 Schematics of trapping-induced hysteresis and ambipolar-to-unipolar transition. a) Schematics of the transfer curves at low temperature and room temperature. Insets are schematics of dominant carriers trapping in perovskite trap states, which generates hysteresis. At room temperature, the increased carrier velocity eliminates the effective trapping states. b) Trapping (p-type trapping state) of mobile carriers close to or in the channel (dominant at low temperatures) causes the hysteresis; above the phase transition temperature (160 K) of the perovskite film, the transistor behavior changes from p-type to ambipolar. At the same time, trapping is weaker at higher temperatures, which suppresses hysteresis.

The temperature-dependent TFT transfer characteristics provide valuable insights into the mechanism of carrier transport in the PVK/SWCNT heterostructure. As shown in the schematics in **Figure 6.17a** (data in **Figure 6.14**), there are two key features of the transfer curves: one is that the channel shows p-type behavior at low temperature and ambipolar behavior at room temperature; the other is that the hysteresis at low temperature is much more pronounced than that at room temperature. These observations suggest that the temperature-dependent characteristics of charge traps play a key role in channel transport.

According to the SRH (Shockley-Read-Hall) model (**Figure 6.17b**), carriers from the valence band/conduction band are captured by trap states at a rate proportional to the capture cross-section multiplied by the thermal velocity of the carriers, or $P_{capture} \propto \sqrt{T}$ [63]. In addition, based on the Heuristic interpretation of the SRH model, considering the different occupation probabilities of the energy levels, the probability of emission is raised by the Boltzmann factor, or $P_{emission} \propto \sqrt{T} exp\left(-\frac{E_a}{kT}\right)$. As a result, charge capture dominates at low temperatures, but the probability of emission increases faster with increasing temperature [58]. The p-type transfer curves at low temperatures indicate that electrons are captured by trap states and the channel transport is dominated by holes. However, more electrons are released from traps with increasing temperature, leading to the ambipolar nature of perovskite transistors at room temperature. Our data imply that hole trap states are located sufficiently close to the valence band edge so that holes can be effectively emitted into the valence band at low temperatures, leading to the observed p-type transport. In contrast, since electron trap states are located deeper in the bandgap, negligible concentrations of electrons are released to the conduction band at low temperatures. At room temperature, the population of trap states are reduced because of the higher kinetic energy of electrons and higher probability of emission, which leads to the ambipolar behavior.

6.3 Conclusion

In summary, semiconducting-enriched single-walled carbon nanotubes (95% s-SWCNT) have been used to enhance the performance of mixed-cation perovskite transistors and inverters, enabling solution-processed mixed-dimensional electronics with high mobility and low operation bias. The mixed-dimensional PVK/SWCNT heterostructures exhibited a low OFF current (10^{-11} A), leading to an ON/OFF ratio up to 10^7, which is the highest reported to date for solution-processed devices. Furthermore, we fabricated high-gain inverters with a switching voltage of 1 V, which provides opportunities for digital electronic applications. Finally, our temperature-dependent transport study revealed a transition from high-temperature ambipolar to low-temperature unipolar behavior, which is a result of competing carrier trapping and emission. The demonstrated merits of low-cost solution processing, high carrier mobility, low-voltage operation, high ON/OFF ratio, and versatile operation modes make mixed-dimensional PVK/SWCNT heterostructures a highly promising candidate for a variety of electronic applications.

6.4 Experimental Section

6.4.1 SWCNT film preparation

Unsorted (80% semiconducting purity) SWCNTs were dispersed by sonicating 75 mg HiPCO (NanoIntegris) powder, 10 wt% SWCNTs and 90 wt% binder, in 15 ml 1% w/v sodium dodecyl sulfate in deionized water by a 0.125 inch tip horn ultrasonicator at 13 W for 60 min, cooled by an ice water bath. The resulting solution was then centrifuged at 10,400*g* for 15 minutes, to separate the individual SWCNTs in the supernatant from the SWCNT bundles, carbonaceous impurities, and metal catalyst particles in the pellet. The supernatant was vacuum filtered on a Nylon membrane and rinsed with copious amounts of ethanol, acetone, and isopropanol to remove residual surfactant. Films were then air dried, peeled off the nylon membrane, and placed in vials for later processing. For sorted (95% semiconducting purity) semiconducting single-walled carbon nanotubes (s-SWCNTs), 150 mg HiPCo powder was similarly dispersed in an aqueous solution of 1% w/v sodium cholate and 25.6% w/v iodixanol for the same power and duration as the unsorted solution. After centrifuging the sonicated solutions at 10,400*g* for 15 min, the supernatant was subsequently sorted by electronic type using density gradient ultracentrifugation and characterized for electronic-type purity as previously reported [43]. Purified fractions were combined into one solution, which was then vacuum filtered and stored as stated above.

6.4.2 PVK/SWCNT precursor preparation

The SWCNTs were dissolved in DMF (Sigma-Aldrich, 99.8%) and DMSO (Sigma-Aldrich, 99.7%) solvent (DMF/DMSO = 4:1 volume ratio) after which the dispersion solvent was ultra-sonicated for 8 hours and then added to the mixed compounds. The perovskite precursor solution contains 0.8 M Methylammonium iodide (MAI, Dyesol), 0.8 M lead (II) iodide (PbI_2, Sigma-Aldrich, 98%), 0.08 M formamidinium bromide (FABr, Dyesol), and 0.08 M lead (II) bromide ($PbBr_2$, Sigma-Aldrich, 98%). The mixed precursor was ultra-sonicated for 2 hours before deposition.

6.4.3 Fabrication and characterization of TFT devices

Heavily doped Si substrates were cleaned with hydrofluoric acid (HF) and ultrasonicated in deionized water, ethanol, and acetone, respectively. After drying in flowing nitrogen, a thin layer (50 nm) of HfO_2 was grown by atomic layer deposition. The Si/HfO_2 substrates were treated with oxygen plasma before film deposition. Au (80 nm) source (S) and drain (D) electrodes were deposited by thermal evaporation through a shadow mask, defining a channel with length of 50 μm and width of 1000 μm. A PEIE solution (0.4 w.t.% in 2-methoxyethanol) was spin coated on the electrodes at 5000 rpm for 30 s. The substrates were then heated at 100 °C for 10 minutes, and followed with ultrasonication in DI water for 10 minutes. A thin layer of PEIE (~5 nm) was deposited after drying the substrates in nitrogen and annealing at 100 °C for 5 minutes in a glovebox. The mixed solution was then spin-coated on the substrates at 2000 rpm for 10 s. A two-step spin coating was performed at 6000 rpm for 30 s, followed by 50 μL of anti-solvent (toluene) dripping during the last 10 s, after which the devices were annealed at 100 °C for 1 hour in a glove box to reduce the charge traps and improve the Ohmic contact between the active layer and electrodes. For comparison, the pristine perovskite films were prepared without SWCNTs following the same protocol.

References

1. Jeon, N. J.; Na, H.; Jung, E. H.; Yang, T. Y.; Lee, Y. G.; Kim, G.; Shin, H. W.; Seok, S. I.; Lee, J.; Seo, J., A fluorene-terminated hole-transporting material for highly efficient and stable perovskite solar cells. *Nat. Energy* **2018,** *3* (8), 682-689.

2. Wang, H.; Kim, D. H., Perovskite-based photodetectors: materials and devices. *Chem. Soc. Rev.* **2017,** *46* (17), 5204-5236.

3. Dou, L.; Yang, Y. M.; You, J.; Hong, Z.; Chang, W. H.; Li, G.; Yang, Y., Solution-processed hybrid perovskite photodetectors with high detectivity. *Nat. Commun.* **2014,** *5*, 5404.

4. Ma, C.; Shi, Y. M.; Hu, W. J.; Chiu, M. H.; Liu, Z. X.; Bera, A.; Li, F.; Wang, H.; Li, L. J.; Wu, T., Heterostructured WS2/CH3NH3PbI3 Photoconductors with Suppressed Dark Current and Enhanced Photodetectivity. *Adv. Mater.* **2016,** *28* (19), 3683-3689.

5. Fang, Y. J.; Huang, J. S., Resolving Weak Light of Sub-picowatt per Square Centimeter by Hybrid Perovskite Photodetectors Enabled by Noise Reduction. *Adv. Mater.* **2015,** *27* (17), 2804-2810.

6. Ling, Y.; Yuan, Z.; Tian, Y.; Wang, X.; Wang, J. C.; Xin, Y.; Hanson, K.; Ma, B.; Gao, H., Bright Light-Emitting Diodes Based on Organometal Halide Perovskite Nanoplatelets. *Adv. Mater.* **2016,** *28* (2), 305-11.

7. Li, F.; Wang, H.; Kufer, D.; Liang, L. L.; Yu, W. L.; Alarousu, E.; Ma, C.; Li, Y. Y.; Liu, Z. X.; Liu, C. X.; Wei, N. N.; Wang, F.; Chen, L.; Mohammed, O. F.; Fratalocchi, A.; Liu, X. G.; Konstantatos, G.; Wu, T., Ultrahigh Carrier Mobility Achieved in Photoresponsive Hybrid Perovskite Films via Coupling with Single-Walled Carbon Nanotubes. *Adv. Mater.* **2017,** *29* (16), 1602432.

8. Senanayak, S. P.; Yang, B. Y.; Thomas, T. H.; Giesbrecht, N.; Huang, W. C.; Gann, E.; Nair, B.; Goedel, K.; Guha, S.; Moya, X.; McNeill, C. R.; Docampo, P.; Sadhanala, A.; Friend, R. H.; Sirringhaus, H., Understanding charge transport in lead iodide perovskite thin-film field-effect transistors. *Sci. Adv.* **2017,** *3* (1), e1601935.

9. Wang, G. M.; Li, D. H.; Cheng, H. C.; Li, Y. J.; Chen, C. Y.; Yin, A. X.; Zhao, Z. P.; Lin, Z. Y.; Wu, H.; He, Q. Y.; Ding, M. N.; Liu, Y.; Huang, Y.; Duan, X. F., Wafer-scale growth of large arrays of perovskite microplate crystals for functional electronics and optoelectronics. *Sci. Adv.* **2015,** *1* (9), e1500613.

10. Li, F.; Ma, C.; Wang, H.; Hu, W. J.; Yu, W. L.; Sheikh, A. D.; Wu, T., Ambipolar solution-processed hybrid perovskite phototransistors. *Nat. Commun.* **2015,** *6, 8238.*

11. Kim, Y. C.; Kim, K. H.; Son, D. Y.; Jeong, D. N.; Seo, J. Y.; Choi, Y. S.; Han, I. T.; Lee, S. Y.; Park, N. G., Printable organometallic perovskite enables large-area, low-dose X-ray imaging. *Nature* **2017,** *550* (7674), 87-91.

12. Shrestha, S.; Fischer, R.; Matt, G. J.; Feldner, P.; Michel, T.; Osvet, A.; Levchuk, I.; Merle, B.; Golkar, S.; Chen, H. W.; Tedde, S. F.; Schmidt, O.; Hock, R.; Ruhrig, M.; Goken, M.; Heiss, W.; Anton, G.; Brabec, C. J., High-performance direct conversion X-ray detectors based on sintered hybrid lead triiodide perovskite wafers. *Nat. Photonics* **2017,** *11* (7), 436-440.

13. Palazon, F.; Akkerman, Q. A.; Prato, M.; Manna, L., X-ray Lithography on Perovskite Nanocrystals Films: From Patterning with Anion-Exchange Reactions to Enhanced Stability in Air and Water. *ACS Nano* **2016,** *10* (1), 1224-1230.
14. Yakunin, S.; Dirin, D. N.; Shynkarenko, Y.; Morad, V.; Cherniukh, I.; Nazarenko, O.; Kreil, D.; Nauser, T.; Kovalenko, M. V., Detection of gamma photons using solution-grown single crystals of hybrid lead halide perovskites. *Nat. Photonics* **2016,** *10* (9), 585-589.
15. Wei, H. T.; DeSantis, D.; Wei, W.; Deng, Y. H.; Guo, D. Y.; Savenije, T. J.; Cao, L.; Huang, J. S., Dopant compensation in alloyed $CH_3NH_3PbBr_{3-x}Cl_x$ perovskite single crystals for gamma-ray spectroscopy. *Nat. Mater.* **2017,** *16* (8), 826-833.
16. Herz, L. M., Charge-Carrier Mobilities in Metal Halide Perovskites: Fundamental Mechanisms and Limits. *ACS Energy Lett.* **2017,** *2* (7), 1539-1548.
17. Johnston, M. B.; Herz, L. M., Hybrid Perovskites for Photovoltaics: Charge-Carrier Recombination, Diffusion, and Radiative Efficiencies. *Acc. Chem. Res.* **2016,** *49* (1), 146-154.
18. Xing, G. C.; Wu, B.; Wu, X. Y.; Li, M. J.; Du, B.; Wei, Q.; Guo, J.; Yeow, E. K. L.; Sum, T. C.; Huang, W., Transcending the slow bimolecular recombination in lead-halide perovskites for electroluminescence. *Nat. Commun.* **2017,** *8*, 14558.
19. Adinolfi, V.; Yuan, M.; Comin, R.; Thibau, E. S.; Shi, D.; Saidaminov, M. I.; Kanjanaboos, P.; Kopilovic, D.; Hoogland, S.; Lu, Z. H.; Bakr, O. M.; Sargent, E. H., The In-Gap Electronic State Spectrum of Methylammonium Lead Iodide Single-Crystal Perovskites. *Adv. Mater.* **2016,** *28* (17), 3406-3410.
20. Shi, D.; Adinolfi, V.; Comin, R.; Yuan, M. J.; Alarousu, E.; Buin, A.; Chen, Y.; Hoogland, S.; Rothenberger, A.; Katsiev, K.; Losovyj, Y.; Zhang, X.; Dowben, P. A.; Mohammed, O. F.; Sargent, E. H.; Bakr, O. M., Low trap-state density and long carrier diffusion in organolead trihalide perovskite single crystals. *Science* **2015,** *347* (6221), 519-522.
21. Richter, J. M.; Abdi-Jalebi, M.; Sadhanala, A.; Tabachnyk, M.; Rivett, J. P. H.; Pazos-Outon, L. M.; Godel, K. C.; Price, M.; Deschler, F.; Friend, R. H., Enhancing photoluminescence yields in lead halide perovskites by photon recycling and light out-coupling. *Nat. Commun.* **2016,** *7*, 13941.
22. Pazos-Outon, L. M.; Szumilo, M.; Lamboll, R.; Richter, J. M.; Crespo-Quesada, M.; Abdi-Jalebi, M.; Beeson, H. J.; Vrucinic, M.; Alsari, M.; Snaith, H. J.; Ehrler, B.; Friend, R. H.; Deschler, F., Photon recycling in lead iodide perovskite solar cells. *Science* **2016,** *351* (6280), 1430-1433.
23. Stranks, S. D.; Eperon, G. E.; Grancini, G.; Menelaou, C.; Alcocer, M. J. P.; Leijtens, T.; Herz, L. M.; Petrozza, A.; Snaith, H. J., Electron-Hole Diffusion Lengths Exceeding 1 Micrometer in an Organometal Trihalide Perovskite Absorber. *Science* **2013,** *342* (6156), 341-344.
24. Kagan, C. R.; Mitzi, D. B.; Dimitrakopoulos, C. D., Organic-inorganic hybrid materials as semiconducting channels in thin-film field-effect transistors. *Science* **1999,** *286* (5441), 945-947.

25. Yusoff, A. B.; Kim, H. P.; Li, X.; Kim, J.; Jang, J.; Nazeeruddin, M. K., Ambipolar Triple Cation Perovskite Field Effect Transistors and Inverters. *Adv. Mater.* **2017,** *29* (8), 1062940.
26. Cheng, H. C.; Wang, G. M.; Li, D. H.; He, Q. Y.; Yin, A. X.; Liu, Y.; Wu, H.; Ding, M. N.; Huang, Y.; Duan, X. F., van der Waals Heterojunction Devices Based on Organohalide Perovskites and Two-Dimensional Materials. *Nano Lett.* **2016,** *16* (1), 367-373.
27. Lee, Y.; Kwon, J.; Hwang, E.; Ra, C. H.; Yoo, W. J.; Ahn, J. H.; Park, J. H.; Cho, J. H., High-Performance Perovskite-Graphene Hybrid Photodetector. *Adv. Mater.* **2015,** *27* (1), 41-46.
28. Chang, P. H.; Liu, S. Y.; Lan, Y. B.; Tsai, Y. C.; You, X. Q.; Li, C. S.; Huang, K. Y.; Chou, A. S.; Cheng, T. C.; Wang, J. K.; Wu, C. I., Ultrahigh Responsivity and Detectivity Graphene-Perovskite Hybrid Phototransistors by Sequential Vapor Deposition. *Sci. Rep.* **2017,** *7*, 46281.
29. De Volder, M. F. L.; Tawfick, S. H.; Baughman, R. H.; Hart, A. J., Carbon Nanotubes: Present and Future Commercial Applications. *Science* **2013,** *339* (6119), 535-539.
30. Javey, A.; Guo, J.; Wang, Q.; Lundstrom, M.; Dai, H. J., Ballistic carbon nanotube field-effect transistors. *Nature* **2003,** *424* (6949), 654-657.
31. Xie, Y.; Gong, M. G.; Shastry, T. A.; Lohrman, J.; Hersam, M. C.; Ren, S. Q., Broad-Spectral-Response Nanocarbon Bulk-Heterojunction Excitonic Photodetectors. *Adv. Mater.* **2013,** *25* (25), 3433-3437.
32. Durkop, T.; Getty, S. A.; Cobas, E.; Fuhrer, M. S., Extraordinary mobility in semiconducting carbon nanotubes. *Nano Lett.* **2004,** *4* (1), 35-39.
33. Avouris, P.; Freitag, M.; Perebeinos, V., Carbon-nanotube photonics and optoelectronics. *Nat. Photonics* **2008,** *2* (6), 341-350.
34. Ausman, K. D.; Piner, R.; Lourie, O.; Ruoff, R. S.; Korobov, M., Organic solvent dispersions of single-walled carbon nanotubes: Toward solutions of pristine nanotubes. *J. Phys. Chem. B* **2000,** *104* (38), 8911-8915.
35. Yamada, T.; Hayamizu, Y.; Yamamoto, Y.; Yomogida, Y.; Izadi-Najafabadi, A.; Futaba, D. N.; Hata, K., A stretchable carbon nanotube strain sensor for human-motion detection. *Nat. Nanotechnol.* **2011,** *6* (5), 296-301.
36. Ghosh, S.; Bachilo, S. M.; Weisman, R. B., Advanced sorting of single-walled carbon nanotubes by nonlinear density-gradient ultracentrifugation. *Nat. Nanotechnol.* **2010,** *5* (6), 443-450.
37. Baughman, R. H.; Zakhidov, A. A.; de Heer, W. A., Carbon nanotubes - the route toward applications. *Science* **2002,** *297* (5582), 787-792.
38. Matsuda, Y.; Tahir-Kheli, J.; Goddard, W. A., Definitive Band Gaps for Single-Wall Carbon Nanotubes. *J. Phys. Chem. Lett.* **2010,** *1* (19), 2946-2950.
39. Schulz, P.; Dowgiallo, A. M.; Yang, M. J.; Zhu, K.; Blackburn, J. L.; Berry, J. J., Charge Transfer Dynamics between Carbon Nanotubes and Hybrid Organic Metal Halide Perovskite Films. *J. Phys. Chem. Lett.* **2016,** *7* (3), 418-425.
40. Blackburn, J. L., Semiconducting Single-Walled Carbon Nanotubes in Solar Energy Harvesting. *ACS Energy Lett.* **2017,** *2* (7), 1598-1613.

41. Arnold, M. S.; Green, A. A.; Hulvat, J. F.; Stupp, S. I.; Hersam, M. C., Sorting carbon nanotubes by electronic structure using density differentiation. *Nat. Nanotechnol.* **2006,** *1* (1), 60-65.
42. Flavel, B. S.; Moore, K. E.; Pfohl, M.; Kappes, M. M.; Hennrich, F., Separation of Single-Walled Carbon Nanotubes with a Gel Permeation Chromatography System. *ACS Nano* **2014,** *8* (9), 9687-9687.
43. Green, A. A.; Hersam, M. C., Nearly Single-Chirality Single-Walled Carbon Nanotubes Produced via Orthogonal Iterative Density Gradient Ultracentrifugation. *Adv. Mater.* **2011,** *23* (19), 2185-2189.
44. Nish, A.; Hwang, J. Y.; Doig, J.; Nicholas, R. J., Highly selective dispersion of single-walled carbon nanotubes using aromatic polymers. *Nat. Nanotechnol.* **2007,** *2* (10), 640-6.
45. Jariwala, D.; Marks, T. J.; Hersam, M. C., Mixed-dimensional van der Waals heterostructures. *Nat. Mater.* **2017,** *16* (2), 170-181.
46. Jeon, N. J.; Noh, J. H.; Kim, Y. C.; Yang, W. S.; Ryu, S.; Seok, S. I., Solvent engineering for high-performance inorganic-organic hybrid perovskite solar cells. *Nat. Mater.* **2014,** *13* (9), 897-903.
47. Jeon, N. J.; Noh, J. H.; Yang, W. S.; Kim, Y. C.; Ryu, S.; Seo, J.; Seok, S. I., Compositional engineering of perovskite materials for high-performance solar cells. *Nature* **2015,** *517* (7535), 476-480.
48. Saliba, M.; Matsui, T.; Domanski, K.; Seo, J. Y.; Ummadisingu, A.; Zakeeruddin, S. M.; Correa-Baena, J. P.; Tress, W. R.; Abate, A.; Hagfeldt, A.; Gratzel, M., Incorporation of rubidium cations into perovskite solar cells improves photovoltaic performance. *Science* **2016,** *354* (6309), 206-209.
49. Salado, M.; Calio, L.; Berger, R.; Kazim, S.; Ahmad, S., Influence of the mixed organic cation ratio in lead iodide based perovskite on the performance of solar cells. *Phys. Chem. Chem. Phys.* **2016,** *18* (39), 27148-27157.
50. Pramanik, C.; Gissinger, J. R.; Kumar, S.; Heinz, H., Carbon Nanotube Dispersion in Solvents and Polymer Solutions: Mechanisms, Assembly, and Preferences. *ACS Nano* **2017,** *11* (12), 12805-12816.
51. Wang, H.; Yu, L. L.; Lee, Y. H.; Shi, Y. M.; Hsu, A.; Chin, M. L.; Li, L. J.; Dubey, M.; Kong, J.; Palacios, T., Integrated Circuits Based on Bilayer MoS_2 Transistors. *Nano Lett.* **2012,** *12* (9), 4674-4680.
52. Kundu, S.; Wang, Y. M.; Xia, W.; Muhler, M., Thermal Stability and Reducibility of Oxygen-Containing Functional Groups on Multiwalled Carbon Nanotube Surfaces: A Quantitative High-Resolution XPS and TPD/TPR Study. *J. Phys. Chem. C* **2008,** *112* (43), 16869-16878.
53. Bai, Y.; Dong, Q. F.; Shao, Y. C.; Deng, Y. H.; Wang, Q.; Shen, L.; Wang, D.; Wei, W.; Huang, J. S., Enhancing stability and efficiency of perovskite solar cells with crosslinkable silane-functionalized and doped fullerene. *Nat. Commun.* **2016,** *7*, 12806.
54. Chen, Q.; Zhou, H. P.; Song, T. B.; Luo, S.; Hong, Z. R.; Duan, H. S.; Dou, L. T.; Liu, Y. S.; Yang, Y., Controllable Self-Induced Passivation of Hybrid Lead Iodide Perovskites toward High Performance Solar Cells. *Nano Lett.* **2014,** *14* (7), 4158-4163.

55. Yang, M. J.; Zeng, Y. N.; Li, Z.; Kim, D. H.; Jiang, C. S.; van de Lagemaat, J.; Zhu, K., Do grain boundaries dominate non-radiative recombination in $CH_3NH_3PbI_3$ perovskite thin films? *Phys. Chem. Chem. Phys.* **2017,** *19* (7), 5043-5050.

56. Li, D. H.; Cheng, H. C.; Wang, Y. L.; Zhao, Z. P.; Wang, G. M.; Wu, H.; He, Q. Y.; Huang, Y.; Duan, X. F., The Effect of Thermal Annealing on Charge Transport in Organolead Halide Perovskite Microplate Field-Effect Transistors. *Adv. Mater.* **2017,** *29* (4), 1601959.

57. Wu, K. W.; Bera, A.; Ma, C.; Du, Y. M.; Yang, Y.; Li, L.; Wu, T., Temperature-dependent excitonic photoluminescence of hybrid organometal halide perovskite films. *Phys. Chem. Chem. Phys.* **2014,** *16* (41), 22476-22481.

58. Park, R. S.; Shulaker, M. M.; Hills, G.; Liyanage, L. S.; Lee, S.; Tang, A.; Mitra, S.; Wong, H. S. P., Hysteresis in Carbon Nanotube Transistors: Measurement and Analysis of Trap Density, Energy Level, and Spatial Distribution. *ACS Nano* **2016,** *10* (4), 4599-4608.

59. Zhang, Z.; Yates, J. T., Band Bending in Semiconductors: Chemical and Physical Consequences at Surfaces and Interfaces. *Chem. Rev.* **2012,** *112* (10), 5520-5551.

60. Zhang, X. N.; Bi, S. Q.; Zhou, J. Y.; You, S.; Zhou, H. Q.; Zhang, Y.; Tang, Z. Y., Temperature-dependent charge transport in solution-processed perovskite solar cells with tunable trap concentration and charge recombination. *J. Mater. Chem. C* **2017,** *5* (36), 9376-9382.

61. Yin, W. J.; Shi, T. T.; Yan, Y. F., Unusual defect physics in CH3NH3PbI3 perovskite solar cell absorber. *Appl. Phys. Lett.* **2014,** *104* (6), 063903.

62. Li, D. H.; Wu, H.; Cheng, H. C.; Wang, G. M.; Huang, Y.; Duan, X. F., Electronic and Ionic Transport Dynamics in Organolead Halide Perovskites. *ACS Nano* **2016,** *10* (7), 6933-6941.

63. Egginger, M.; Bauer, S.; Schwodiauer, R.; Neugebauer, H.; Sariciftci, N. S., Current versus gate voltage hysteresis in organic field effect transistors. *Monatsh Chem.* **2009,** *140* (7), 735-750.

Chapter 7

Multi-Input Parameter Modulable Memtransistors from Hybrid Perovskite/ Conjugated Polymer Heterostructures

Synopsis

In this chapter, we report the development of hybrid memtransistor, modulable by multiple physical inputs (optical and electrical inputs). Neuromorphic computing has the potential to address the inherent limitations of conventional integrated circuit technology, ranging from perception, pattern recognition to memory and decision-making[1-3]. Despite their low power consumption[4], traditional two-terminal memristors can perform only a single function while lacking heterosynaptic plasticity.[5] Inspired by unconditioned reflex, multi-terminal memristive transistors (memtransistor) were developed to realize complex functions, such as multi-terminal modulation and heterosynaptic plasticity[6]. Here we combine a hybrid metal halide perovskite with an organic conjugated polymer to form heterojunction transistors that are responsive to both electrical and optical stimuli. We show that the synergistic effects of photo-induced ion migration in the perovskite and electronic transport in the polymer layers can be exploited to realize multiple-stimulus memristive functions. The device combines reversible, non-volatile conductance modulation with large switching current ratios, high endurance and long retention times. Using *in-situ* scanning Kelvin probe microscopy and variable temperature charge transport measurement, we correlate the collective effects of bias- and photo-induced ion migration with the heterosynaptic behaviour observed in this bio-inspired memtransistor.

7.1 Introduction

Inspired by the morphology and function of 10^{14} synapses in the mammalian brain, silicon-based asynchronous spiking neural networks (SNNs) have ushered in an era of developing non-volatile resistive switching as memristors for artificial intelligence[7]. However, several attempts to break through the Von Neumann bottleneck, such as crossbar neural network and field-effect transistors with ionic gates[8,9], have failed to demonstrate heterosynaptic plasticity and focused primarily on singular electrical input[6]. Recently, researchers utilized different types of materials to realize neuromorphic computing with tunable resistance switching. Yoeri *et al.* demonstrated an electrochemical neuromorphic organic device based on PEDOT:PSS/PEI[10] while Vinod, *et al.* reported multi-terminal memtransistors from transition metal dichalcogenides (TMDs)[6]. They both utilized the nature of ionic charge transfer and electromigration within the active materials. An alternative approach was adopted by Choi *et al.* which demonstrated epitaxial random-access memory (RAM) devices using etched single-crystalline SiGe in which an ionic pathway was engineered to realize localized filamentary conduction[11]. Overall, various strategies related to resistance switching have been developed, and the exploration of one or more materials with complementary functionalities is becoming the key for future developments.

Over the past decade, methylammonium lead halide perovskites (PVKs) have attracted enormous attention for application in photovoltaic, (opto)electronic and photonic devices due to their intriguing physical properties[12]. In addition to their well-known application in photovoltaics[13,14], PVKs have been explored for other devices such as photodetectors[15], light-emitting diodes and transistors[16,17], and resistance switching devices[18]. An intriguing effect associated with the prototypical PVK material methylammonium lead triiodide ($MAPbI_3$), is the giant switchable photovoltaic effect that results in commonly observed hysteresis in photovoltaic cells[19]. Recent studies have attributed the observed hysteresis to ion migration and the light tunable electron-ion interaction within the halide perovskites[20-22]. This unique mixed conduction character of PVKs inspired us to construct a novel form of non-Von Neumann memory devices whose functions can be tuned by multiple input stimuli to realize neuromorphic computing.

Herein, we describe the development a multi-input parameter modulable memtransistor based on a solution-processed heterojunction (HJ) consisting of the mixed halides perovskite $MAPbI_{3-x}Br_x$ and the indacenodithiophene–benzothiadiazole copolymer (IDT-BT). The memtransistors are able to process different types of input stimuli, such as light and electric signals, due to the modulable Schottky barrier height, present at the injecting electrode interface, by the applied electric-field and light-controlled ion migration in the channel. The devices show high gate tunability ($>10^5$) with large switching current ratios (10^2), high endurance ($>10^2$ times) and excellent retention (10^3 s). Due to the photo-induced halide redistribution occurring during electric field application, the memtransistor exhibit unique light-induced plasticity, making it promising for pattern recognition directly in the optical domain[23]. In-situ scanning Kelvin probe microscopy (SKPM) and temperature dependent charge transport measurements, provide unique insights in the kinetics of the electric bias and photo-induced ion migration that underpin device operation. The seamless integration of optical and electric inputs is the first step towards complex neuromorphic learning.

7.2 Results and Discussions

7.2.1 Optical and structural properties of the hybrid channel

We employed the bottom-contact, top-gate (BC-TG) transistor architecture shown in **Figure 7.1a** to construct the multi-input modulable memtransistors. The mixed halides perovskite (PVK) layer was processed via spin-coating on top of pre-patterned gold (Au) source and drain (S-D) electrodes, followed by spin-coating of the semiconducting polymer IDT-BT. The heterojunction exhibits a smoother surface topography as compared to the pristine PVK layer (**Figure 7.1b**). **Figure 7.1c** shows the energy levels band diagram of the IDT-BT/PVK heterojunction that was constructed using a different experimental technique including, photoelectron spectroscopy (PESA) in air and UV-Vis absorption (**Figure 7.1d**) spectroscopy (**Table 7.1**).

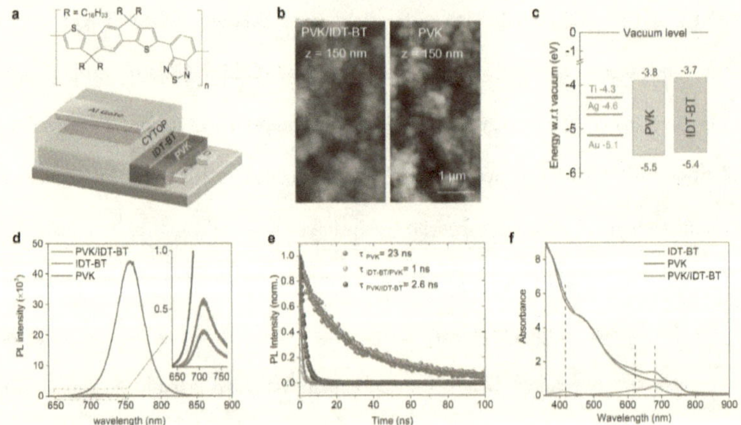

Figure 7.1 Architecture and energy transfer of the IDT-BT/PVK memtransistor. a, Schematic of IDT-BT/PVK memtransistor with top-gate, bottom-contact configuration, in which source/drain electrodes are gold, defining the channel length (L) = 30 μm, width (W)

= 1000 μm, and gate electrode is aluminium. The dielectric layer is CYTOP. The depicted molecular structure represents the IDT-BT copolymer. **b**, AFM images of the IDT-BT/PVK bilayer (left) and pristine PVK (right), indicating the good coverage of IDT-BT and hinting at its penetration into the PVK film. **c**, Electronic band alignment o with electrodes and corresponding charge transfer. **d**, PL spectra of pristine IDT-BT, PVK and the bilayer. **e**, TR-PL decay dynamics of neat PVK film and IDT-BT/PVK bilayer excited from either side at 532 nm, indicating charge transfer occurs on the nanosecond timescale. **f**, UV-Vis spectra for the pristine IDT-BT, PVK and IDT-BT/PVK films.

Table 7.1 Summary of the physical properties of the IDT-BT polymer.

Polymers	M_n (kDa)/M_w(kDa)	λ_{max}/nm (thin film) [b]	$E_{opt.gap}$ [c]	IP (eV) [d]	EA (eV) [e]
IDT-BT	58.3/95.3	666	1.7	-5.4	-3.7

[a] Solution absorption spectrum (in chlorobenzene). M_n: molar mass averages of the number; M_w: molar mass averages of the weight.

[b] film absorption spectra from pristine film.

[c] obtained from the onset value of absorption in pristine film.

[d] measured by Photo-Electron Spectroscopy in Air (PESA) system.

[e] calculated from $E_{opt.gap}$ and IP.

Cross-sectional analysis of the heterojunction channel using SEM suggest the IDT-BT solution penetrates into the PVK layer (**Figure 7.2a-d**). The infiltrated IDT-BT solution does not appear to affect the integrity of the perovskite layer beneath since the UV-Vis absorption features of PVK and IDT-BT remaining identical in the heterojunction (**Figure 7.1d**). Furthermore, **Figure 7.3a,b** show the phase images of the pristine PVK film and the heterojunction, which demonstrate the pristine PVK film has more rigid surfaces and crystal dislocations than the heterojunction. The height histograms in **Figure 7.3c** shows that the heterojunction has a smoother surface, which may be beneficial for carrier transport.

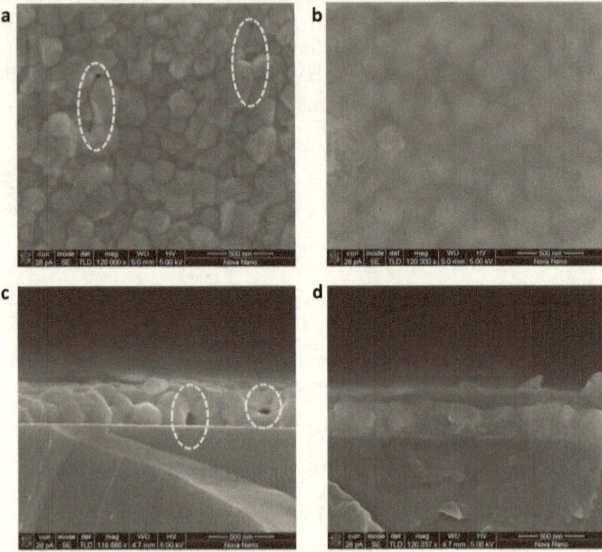

Figure 7.2 a, b. Top-view SEM images of the pristine PVK film and IDT-BT/PVK bilayer. **c, d**. cross-section of the SEM images from the pristine PVK film and IDT-BT/PVK bilayer, respectively. The yellow dash lines circle the pin-holes in the pristine PVK film, while those were filled in by the polymer.

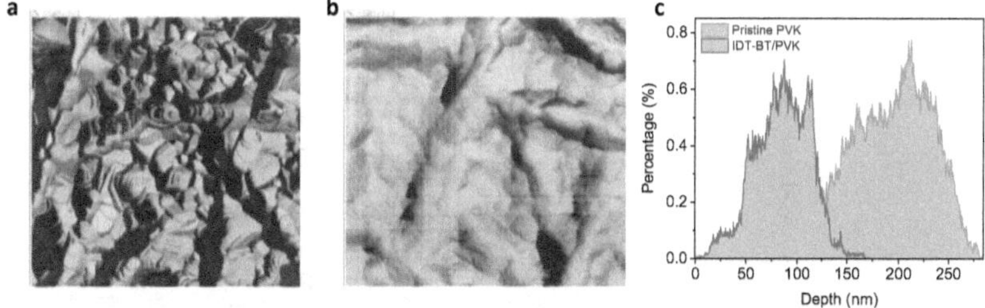

Figure 7.3 a, b. the AFM phase images (2μm × 2μm) of the pristine PVK film and IDT-BT/PVK bilayer, respectively. Pristine PVK film shows more rigid surfaces and crystal dislocation. **c**. histogram of the depth of the pristine PVK film and IDT-BT/PVK bilayer, showing the surface of IDT-BT/PVK bilayer is much smoother.

The formation of type II heterojunction (**Figure 7.1c**) implies the possibility for charge transfer between the two layers upon optical excitation. To test this hypothesis, we performed steady-state photoluminescence (PL) measurements. PVK exhibits a single photoluminescence (PL) peak at 773 nm, which is assigned to radiative carrier recombination (**Figure 7.1d**).[24] The PL signal intensity of the pristine IDT-BT is significantly lower than that of PVK. When IDT-BT is deposited on PVK, the PL signal shows a marked quench with respect to the pristine PVK layer. The UV-Vis absorbance and PL spectra of pristine IDT-BT are shown in **Figure 7.4** for comparison.

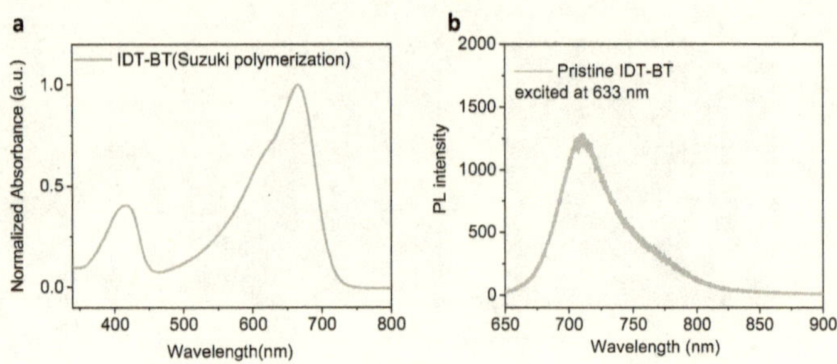

Figure 7.4 the UV-vis spectrum and PL spectrum for pristine IDT-BT film.

The combination of excellent light absorbing and photocarrier generation characteristics of $MAPbI_{3-x}Br_x$[12,25] combined with the narrow energy difference (0.1 eV) between the PVK valence band maximum (VBM) and the highest occupied molecular orbital (HOMO) of IDT-BT (**Figure 7.1c**), supports the charge transfer hypothesis. Further confirmation stems from time-resolved PL (TR-PL) measurements, which yield a PL lifetime of 23 ns of the neat PVK tracked at the emission peak (**Figure 7.1e**). Optical excitation incident to the PVK side of the bilayer PVK/IDT-BT shows a reduced lifetime of 2.6 ns, whereas exciting incident to the IDT-BT side shows an even shorter lifetime of 1 ns. The PL decay of PVK in the bilayer is most likely attributed to diffusion-limited charge extraction and interfacial charge recombination, which are two competing processes. Since the PL lifetime is faster in the latter case, charge transfer from PVK IDT-BT appears to be more efficient upon excitation from the polymer side. The latter process is corroborated by time-dependent evolution of the PL peaks shown in **Figure 7.5**.

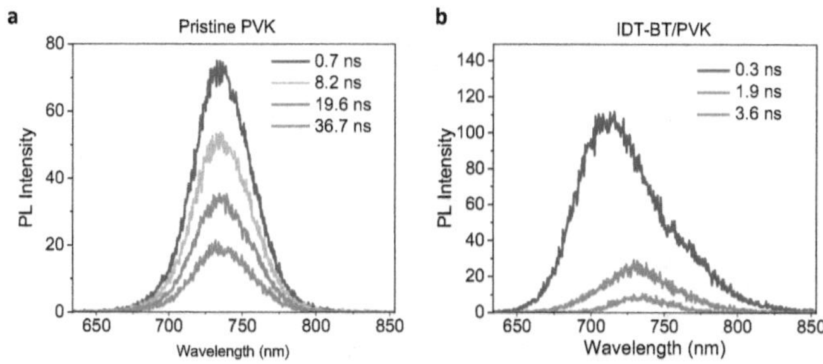

Figure 7.5 Time-dependent evolution of steady PL peaks of the pristine PVK and IDT-BT/PVK bilayer, confirming the energy transfer process occurs in nanosecond scale.

To exclude possible effects of material intermixing, we performed cross-sectional analysis of the PVK/IDT-BT HJs using SEM (**Figure 7.2a-d**). The results show that the IDT-BT solution does not intermix with the PVK layer but appears to fill voids. This observation agrees with the UV-Vis absorption measurements (**Figure 7.1f**) which show that all absorption features associated with the PVK and IDT-BT layers are also preserved in the heterojunction. X-ray diffraction (XRD) measurements also show no difference on the diffraction patterns associated with the individual materials and particularly with the orthorhombic phase of the PVK layer [*i.e.* (110), (220) and (330)] following IDT-BT deposition (**Figure 7.6a**). Similarly, Raman measurements (**Figure 7.6b**) show no change of the fingerprint peaks for IDT-BT upon deposition on PVK.

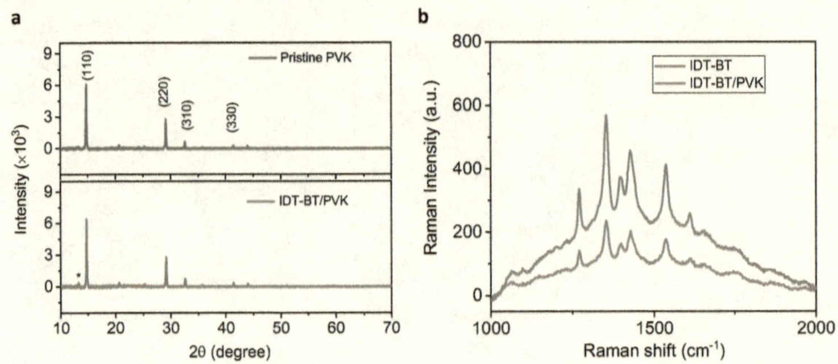

Figure 7.6 a) XRD patterns of the pristine PVK film and IDT-BT/PVK bilayer. Neglectable difference can be observed from that before and after IDT-BT deposition. b) Raman spectra of the pristine PVK film and IDT-BT/PVK bilayer. The fingerprint peaks of IDT-BT have no shift when depositing on PVK layer, showing its stable inherent chemical nature.

7.2.2 Photo-induced carrier transport

The excellent hole transporting characteristics of IDT-BT[26,27] is expected to facilitate the formation of a hole-transporting channel when deposited atop the PVK layer. The TG-BC transistor architecture adopted here (**Figure 7.1a**) is ideal for testing this hypothesis as the channel should form at the interface between the IDT-BT and the CYTOP dielectric. The representative transfer and output characteristics for a pristine IDT-BT based FET are shown in **Figure 7.7**.

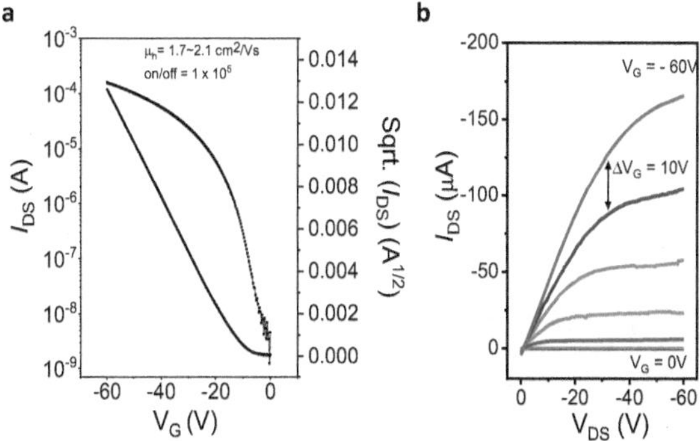

Figure 7.7 A typical transfer (**a**) and output curve (**b**) of pristine IDT-BT, showing a hole mobility of 2.1 cm^2/Vs and an on/off ratio of 10^5.

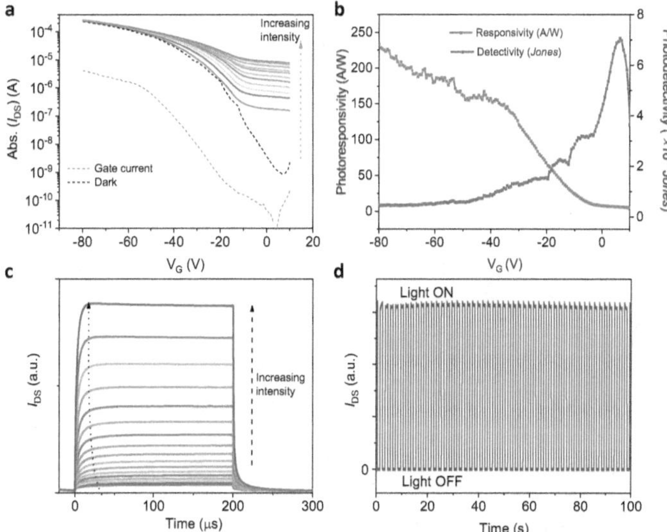

Figure 7.8 Photoresponse of the IDT-BT/PVK memtransistor. a, Transfer characteristics of an IDT-BT/PVK memtransistor under different light intensity (from dark to 7.2 mW/cm^2). The black dash line represents the transfer curve in dark, and the red dash line represents the leakage (I_G) current during the V_G sweeping with V_{DS} = -80 V. **b**,

calculated photoresponsivity and photodetectivity at different V_G and the incident light intensity was fixed at 1 mW/cm^2. The highest photoresponsivity (227 A/W) was achieved at V_G = -80 V, while the highest photodetectivity was achieved at V_G = 6 V. **c**, Transient photocurrent at a light pulse frequency of 2.5 kHz with different white light intensity (from 2 to 53.9 mW/cm^2). **d**, photo-switching characteristics of the device under alternating dark and light illumination (630 nm, 0.21 mW/cm^2).

To test whether the type II heterojunction channel is indeed photoactive, we illuminated the device with different light intensities in the range of 0 to 7.2 mW/cm^2 (**Figure 7.8a**). A systematic increase in the off current of the device is observed with increasing illumination intensity indicating an increased carrier photogeneration. To gain further insight of the charge transport process, we performed transient photocurrent measurements using a square light pulses of 2.5 kHz (53.9 mW/cm^2) and a relatively low source-drain voltage (V_{DS}) of 2 V in order to prevent lateral ion migration which may interfere with the measurements. From the photocurrent response in **Figure 7.8c** we obtain the rise and fall times of ≈28 and ≈46 μs, respectively, yielding a total response time of ≈74 μs. The latter is significantly faster than previously reported data for single-layer perovskite photoconductors, which typically fall in seconds' scale[15,28]. We attribute this to the improved hole-extraction and transport across the hybrid channel.[27,29] Importantly, the photo-switching function of the device remains stable indicating robust operation (**Figure 7.8d**).

7.2.3 Photoinduced ion migration

In addition to charge carrier transport, photoinduced ionic conduction is also anticipated to play an important role within the PVK layer[30]. To better understand its effects, we first studied the crystallinity of the PVK layer at the interfaces and in the bulk using the grazing-incidence wide-angle X-ray scattering (GIWAXS) technique. A two-dimensional GIWAXS pattern of the pristine PVK surface with the incident angle of 0.05° is shown in **Figure 7.9a.**

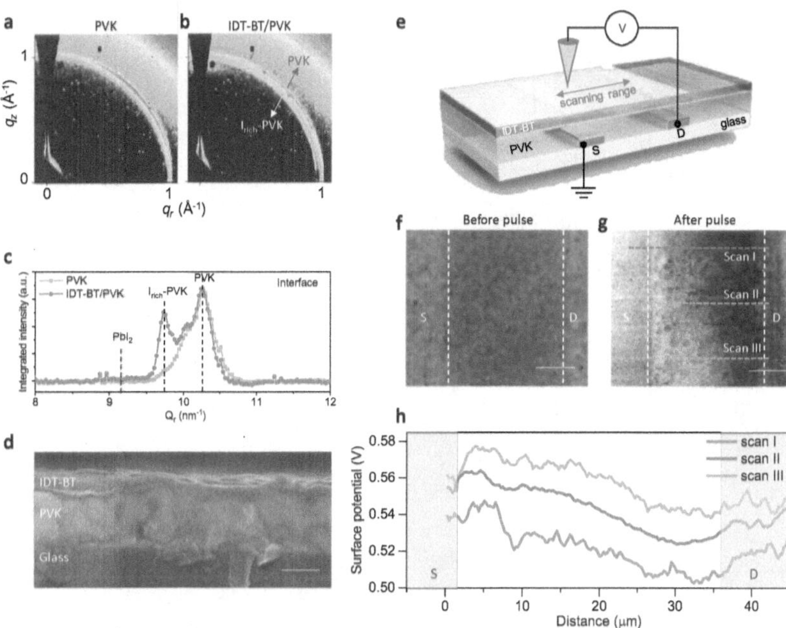

Figure 7.9 Ionic conduction of the IDT-BT/PVK memtransistor. a, 2D-GIWAXS spectra of the pristine PVK surface with an incident angle of 0.05°. **b**, 2D-GIWAXS spectra of the interface of IDT-BT/PVK with the same incident angle. **c**, azimuthally integrated 1D plots showing the comparison of the surface of PVK w/ and w/o IDT-BT layer. **d**, cross-section SEM image of the IDT-BT/PVK stack (scale bar 100 nm). **e**, schematic of the IDT-BT/PVK heterostructure used for scanning Kelvin probe force microscopy (SKPFM). **f, g**, the SKPFM mapping before and 1 hour after the pulse (scale bar 10 μm). **h**, line scan of the surface potential of the heterostructure at three different positions.

The ring at $q_r = 10.27\ nm^{-1}$ is assigned to the (110) face of PVK.[31] The pattern also provides information about the crystallinity of IDT-BT/PVK interface. The diffused ring seen at $q_r = 9.75\ nm^{-1}$ (**Figure 7.9b**) indicates a new phase present at the IDT-BT/PVK interface. A comparison of the azimuthally integrated 1D plots of the surface of PVK w/ and w/o IDT-BT layer is shown in **Figure 7.9c**. Since, the GIWAXS peak for PbI_2 is known to be at $q_r = 9.1\ nm^{-1}$,[32] we ascribe the peak at $q_r = 9.75\ nm^{-1}$ to an iodine-rich perovskite phase due to its larger *d*-spacing. In contrast, a weaker iodine-rich perovskite phase is found in the bulk as evident from the azimuthally integrated 1D plots of the bulk of PVK/IDT-BT shown in **Figure 7.10**.

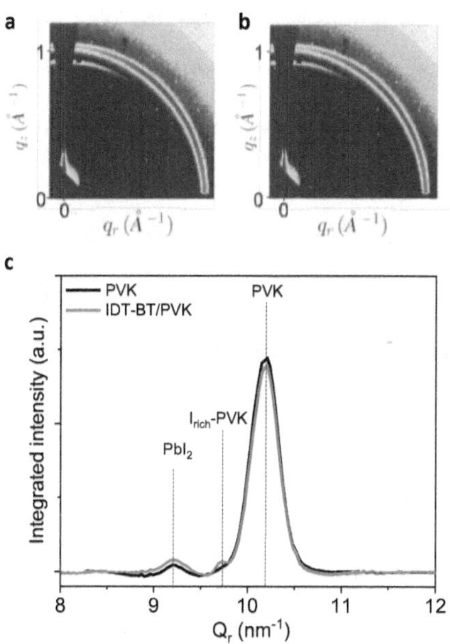

Figure 7.10 2D-GIWAXS spectra of the pristine PVK bulk before (**a**) and after (**b**) IDT-BT deposition with an incident angle of 0.15°. **c.** integrated 1D plots from GIWAXS of the bulk of PVK film and IDT-BT/PVK stack. Iodine-rich phase of perovskite can only be observed in the IDT-BT/PVK stack.

These results suggest that the concentration of I within the PVK layer is higher in regions close to the PVK/IDT-BT interface. Since I is known to be highly mobile within the PVK crystal,[21,30] the presence of I-rich PVK phase present at the interface with IDT-BT could serve as an additional ion pathway, further enhancing the mixed ionic-electronic conduction character of the heterojunction channel.

To directly probe the ion migration and resulting electrostatic perturbations across the heterojunction channel (**Figure 7.9d**), we performed *in situ* scanning Kelvin probe force microscopy (SKPFM) measurements (**Figure 7.9e**). **Figure 7.9f** shows an SKPFM image obtained prior to the application of any bias. Evidently, no surface potential gradient along the channel exists. To trigger ion migration, a voltage pulse (V_{DS} = -20 V) of 10 s duration was applied followed by a SKPFM scan (**Figure 7.9g**). Evidently, the surface potential of the channel closer to the drain electrode increased, indicating the existence of localised ions (I). Line profiles taken from **Figure 7.9g** reveal a potential difference of ≈40 meV between the source-drain electrodes (**Figure 7.9h**), due to the change in the local channel resistance. By comparing the channel's surface potential with its topography (**Figure 7.11**), we identify large potential changes in regions around the grain boundaries. This finding suggests ions segregate mainly at the grain boundaries (interfaces) in agreement with previous results[33].

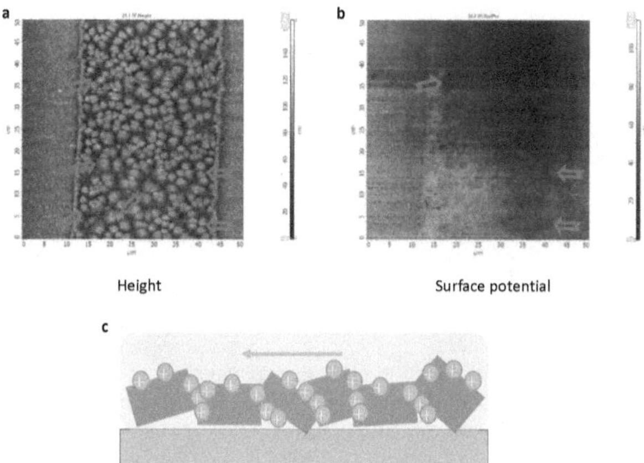

Figure 7.11 a. topography profile of the bilayer in the channel. **b**. surface potential map at the same place. Closely comparing the surface potential map with the topography profile of the channel. **c.** schematics of the ion distribution and migration. We found larger potential changes in the grain boundaries, suggesting the ions migrate mainly in the grain boundaries or at the interface.

The effect of ion migration on the Schottky barrier present at the electrodes/PVK interface, was also studied using temperature-dependent charge transport measurements. Devices were conditioned at V_{DS} = -20 V for 10 s prior to the temperature dependent measurements (**Figure 7.12**). Using the thermionic emission model for the Schottky-barrier FETs (SB-FETs)[6], the effective Schottky barrier heights (ϕ_b) was calculated, yielding values in the range 30-40 meV, in good agreement with the SKPFM measurements in **Figure 7.9f-h**.

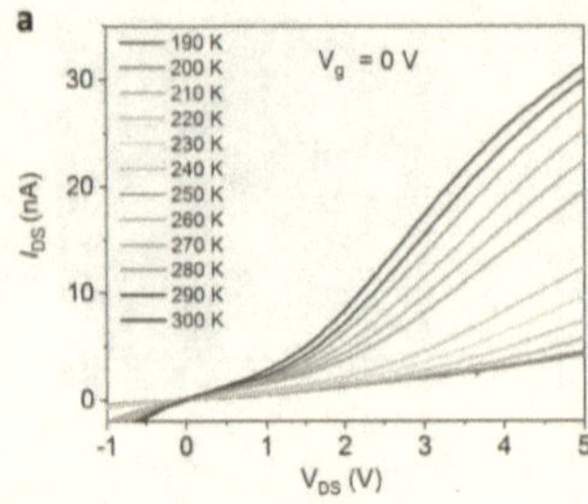

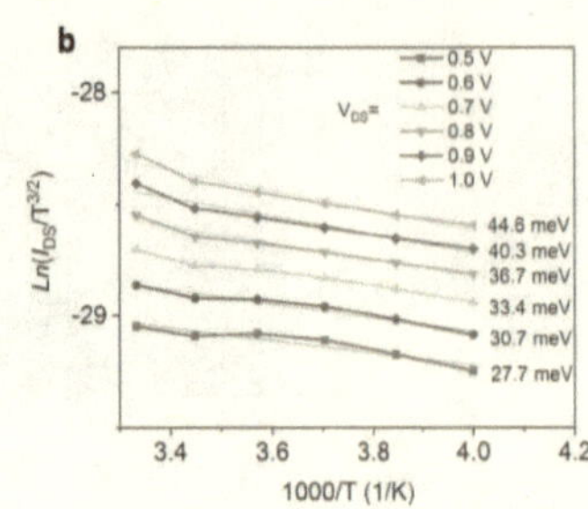

Figure 7.12 a. the variable-temperature transport measurements (I_{DS} $vs.V_{DS}$) from 190 K to 300 K. The ions are frozen under 240 K. **b.** fitting by the thermionic emission model, where the Schottky-barrier FETs (SB-FETs) is in a 2D configuration.

7.2.4 Memtransistor operation

Figure 7.13a shows the electrical characteristics of the heterojunction memtransistor, where a bipolar resistive switching behavior is evident. The PVK/IDT-BT memtransistor enters the high-resistance state (HRS) at the beginning of sweep 1 (60 to 40 V). Applying a few seconds of high-voltage conditioning (training), the device gradually turns to low-resistance state (LRS) and maintains the LRS during the rest of sweep 1 (40 to 0 V). The device turns to the HRS abruptly when the polarity of the source-drain voltage (V_{DS}) is reversed. The latter indicates that the resistive states relates to the carrier injection at the electrode/PVK interface. To verify this hypothesis, we studied devices with the same architecture but different metal electrodes, namely Au, Ti, and Ag. We find the Ag electrodes to be easily etched by the acidic precursor, while the Ti electrodes to be protected by their native oxide (TiO_2). Compared with Au electrodes, transistors with Ti exhibit lower rectification but higher channel conductance (**Figure 7.14**) due to improved carrier injection.

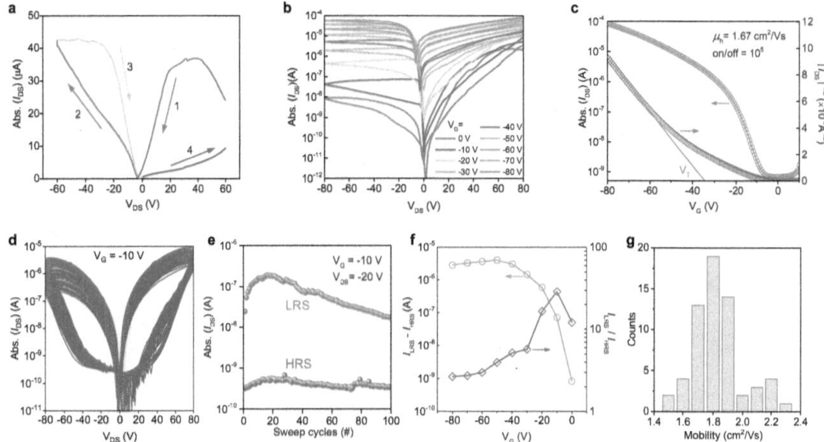

Figure 7.13 Electrical characteristics of the IDT-BT/PVK memtransistor. a, Representative bipolar resistive switching characteristic of an IDT-BT/PVK memtransistor. The sweep direction is 1→4 at V_G = -60 V. **b.** Output curves of the memtransistor showing the programmable levels of switching ratio at different V_G. **c.** A typical transfer curve in the low-resistance state (LRS) (V_{DS} = -80 V), showing mobility of 1.67 cm²/Vs. **d.** 100 sweep cycles of the memtransistor at V_G = -10 V. **e.** Endurance of the current at V_{DS} = -20 V, giving a switching ratio between the low resistance state current (I_{LRS}) and high resistance state current (I_{HRS}) of ~10^2. **f.** current difference (I_{LRS}-I_{HRS}) and switching ratio (I_{LRS}/I_{HRS}) between LRS and HRS at different gate voltage. **g.** Statistical data for the hole mobility of 64 devices, the average mobility ranges between 1.7-1.9 cm²/Vs.

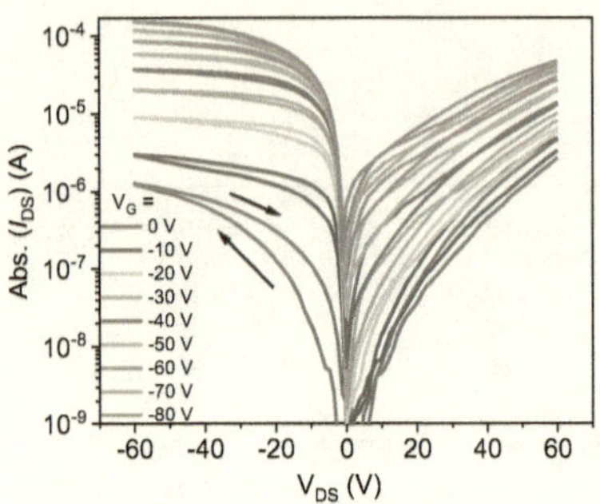

Figure 7.14 A typical scanning sequence of the IDT-BT/PVK memtransistor with titanium electrodes. Output curves of the memtransistor with double linear sweep.

During sweep **2** (0 to -60 V in **Figure 7.13a**), the device gradually switches from HRS to LRS, a state which is maintained during sweep **3** (-60 to 0 V). Again, the channel returns to HRS immediately when the polarity of V_{DS} changes, and the device gradually transits from HRS to LRS during sweep **4** (0 to 60 V). Unlike filamentary resistive switching, our device acts purely as LRS-LRS memtransistor[6]. Importantly, the gate tunability of the channel current is high ($\approx 10^4$) for V_G in the range from 0 to -80 V (**Figure 7.13b**). The operating characteristics of the device are dominated by the current on/off ratio of the polymer channel as well as the PVK layer thickness. To this end, a PVK thickness of $\approx$180 nm (**Figure 7.15**) is found to yield memtransistor with high on/off ratio ($\approx 10^5$) and hole mobility of 1.67 cm^2/Vs (**Figure 3c**). The latter value was calculated in the saturation operating regime using the gradual channel approximation

$$I_{DS} = \mu C_i \frac{W}{2L}(V_G - V_{TH})^2 \quad (7.1)$$

here, C_i is the capacitance of the CYTOP dielectric (ε_r = 2.1 and 900 nm-thick), W and L are the width and length of the channel ($W/L = 1{,}000/30\ \mu m$). And V_{TH} is the threshold voltage.

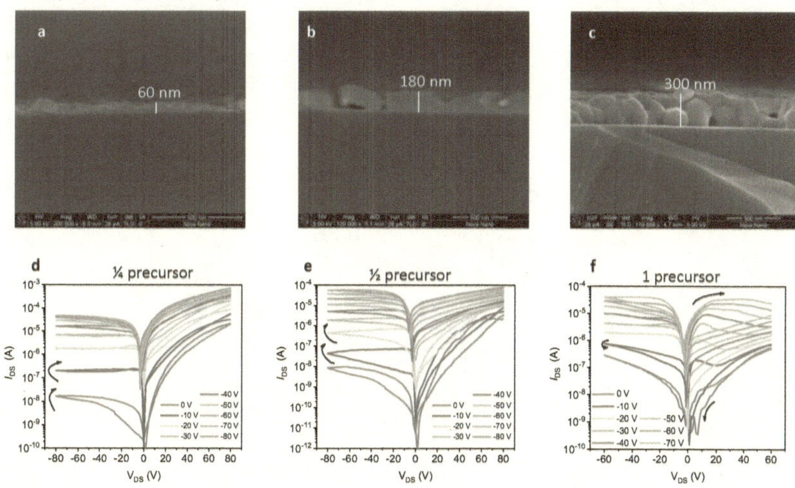

Figure 7.15 a,b,c. cross-section of the SEM images from the pristine PVK film with different concentration of precursors. (0.2M, 0.4M, and 0.8M, respectively). **d,e,f.** corresponding double sweep output curves, showing the thickness of PVK layer is very crucial to the performance of THE memtransistors.

Figure 7.13d displays 100 consecutive I_{DS} *vs.* V_{DS} sweeps measured for an IDT-BT/PVK memtransistor (V_G = -10 V) in the dark, highlighting the highly stable bipolar switching characteristics of the device. **Figure 7.13e** shows the endurance characteristics of the same device while being reprogrammed 100 times between LRS and HRS. The I_{LRS}

increases exponentially during the initial subset of measurement cycles with the I_{HRS} remaining below 1 nA (at V_{DS} = -20 V). The I_{LRS} saturates at ≈10^{-7} A, accompanied by a slow inverse exponential decay. The retention characteristics of the LRS and HRS evaluated within a period of 2 h in the dark, are shown in **Figure 7.16** and project a reliable state storage.

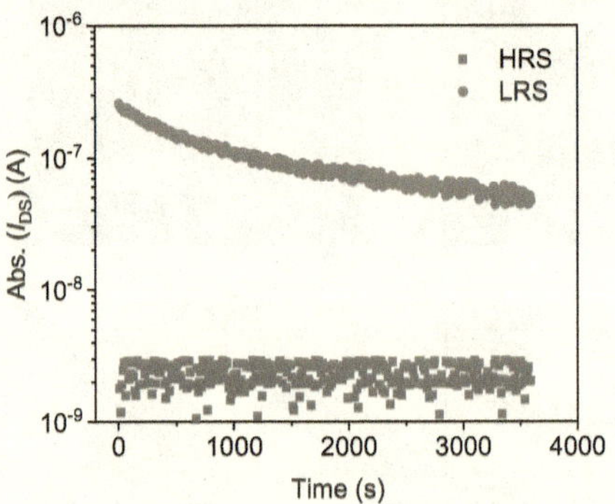

Figure 7.16 retention of the HRS and LRS currents at V_{DS} = -10 V, V_G = -10 V in a 2-hours period in the dark.

In **Figure 7.13f** we plot the current difference (I_{LRS}-I_{HRS}) and switching ratio (I_{LRS}/I_{HRS}) between LRS and HRS. The I_{LRS}-I_{HRS} increases with increasing V_G, the mobile ions. The current switching ratio reaches a peak at V_G = -10 V, followed by a monotonic drop at higher V_G. Importantly, the memristors can be reliably fabricated as evident by the statistical analysis of 64 devices processed during several different runs (**Figure 7.13g**), making them highly promising for neuromorphic computing applications.

Based on these observations we propose a mechanism governing the operation of the hybrid memtransistor. The characteristic resistive switching behaviour can be divided into four regions (**Figure 7.17**), each defined by a certain bias regime.

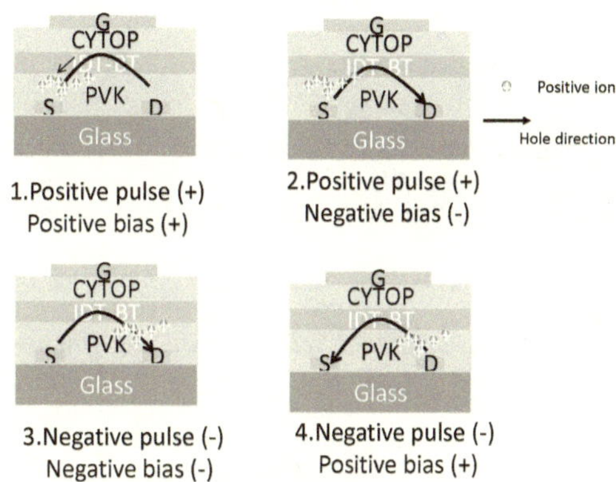

Figure 7.17 schematic showing the ion migration and carrier transport of an IDT-BT/PVK memtransistor, resulting in a bipolar resistive switching behaviour.

In region **1**, positive poling (V_{pol}> 20 V) was applied to drain electrode before measuring the I_{DS} at a positive V_{DS} (>0 V). It has been previously shown that the iodine vacancy ($V_I^{\cdot}$) is the dominant ion in PVK films due to its lower formation energy and higher mobility[34,35]. As such, we consider positive ions ($V_I^{\cdot}$) to be repelled away from the drain electrode (*i.e.* lower concentration of $V_I^{\cdot}$) leading to *n*-type self-doping[19]. The latter will enhance hole injection from Au to PVK due to Schottky barrier height lowering. In region **2**, positive poling remains, however, the I_{DS} is measured at a negative V_{DS} (< 0 V). Here,

PVK layer remains *n*-doped close to the drain terminal, which reduces the I_{DS} and exhibits rectification behaviour similar to a *p-n* junction. In region **3**, the negative poling triggers ion migration with $V_I^{\cdot}$ accumulating close to the drain electrode. The reverse bias changes the character of the bipolar junction from *p-n* to the *n-p* configuration. As a consequence, the I_{DS} is suppressed in region **4**.

To verify the role of V_I, we varied the I concentration within the PVK layer. Since V_I is more mobile than the Br vacancy[36], we partially exchanged I with Br. GIWAXS and device measurements show that PVK layers with higher Br concentration exhibit improved crystallinity but suppressed resistive switching (**Figure 7.18**). On the contrary, PVK devices with higher I content show improved switching behavior but poor field-effect modulation (**Figure 7.18e-g**). Guided by these results, the stoichiometry of the halide concentration was fixed to the optimal ratio of $MAPbI_{2.55}Br_{0.45}$.

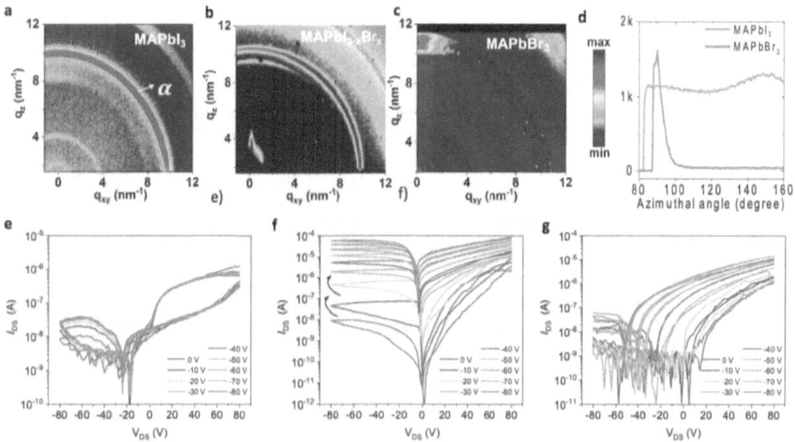

Figure 7.18 a,b,c 2D-GIWAXS spectra of the pristine PVK with different bromide concentration. The spectra were measured with an incident angle of 0.15°, showing the crystalline information of bulk PVK. **d.** The integrated intensities at different azimuthal angle, showing $MAPbBr_3$ has better crystallinity and an out-of-plane growth direction. **e,f,g.** corresponding double sweep output curves, indicating iodine vacancies play key role in the performance.

7.2.5 Multi-input parameter memtransistor operation

Next we studied the influence of light on the electrical characteristics of the memtransistors. Our key hypothesis here is that in addition to the gate field, light should also affect both the charge transport (**Figure 7.8a**) and ion distribution within the channel due to electron-ion interactions in the PVK layer[22] leading to intriguing device behaviour. The aforementioned electron-ion interaction can be described by

$$I_I^x + h^{\cdot} \rightleftharpoons I_i^x + V_I^{\cdot} \tag{7.2}$$

where, I_I^x is neutral iodine atoms on iodine sites, $h^\cdot$ is positive holes, and I_i^x stands for interstitial iodine atoms with a neutral effective charge. Increasing the carrier density (holes) is thus expected to result to a higher V_i. This prediction is in agreement with our experimental observations shown in **Figure 7.13f,** which indicates a higher ion-induced current difference (I_{LRS}-I_{HRS}) with increasing V_G. The data also shows that the concentration of holes ($h^\cdot$) transferred from the PVK layer to IDT-BT depends on both V_G and light leading to a pronounced gate (light)-tunable resistance switching.

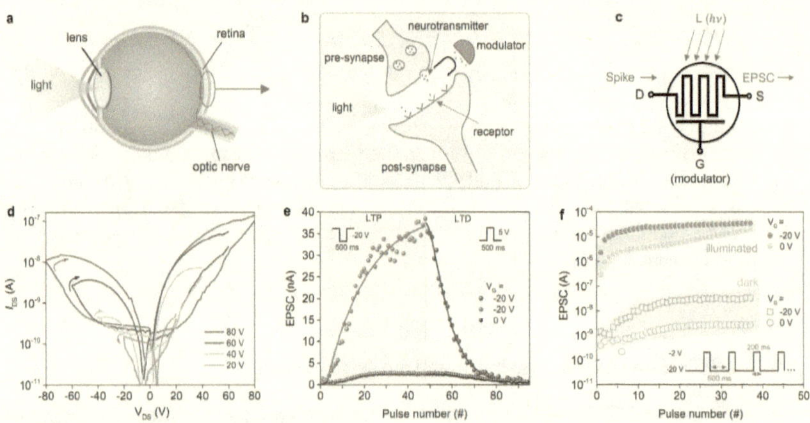

Figure 7.19 Electrical and photonic plasticity of the IDT-BT/PVK memtransistor. a, Schematic illustration of the human visual system. b, schematics of the synapse in the retina. c, electric symbol of the bio-inspired memtransistor. Drain and source terminals imitate the pre- and post-synapse, respectively. Gate terminal imitates the modulator and L symbolizes the light. d, pinched hysteresis loops in dark with different maximum-sweep voltages (V_{DS} from |± 20| V to |± 80| V) at V_G = 0. e, progression plots of the EPSC under dark as a function of the number of input pulse (V_{Pot}=-20 V, V_{Dep}=5 V, t_{width}=500 ms) in dark at the different V_G of 0, and -20 V. f, progression plots of EPSC under light as function of time of input pulses (V_{Pot}=-20 V, t_{width}=500 ms, $t_{interval}$=200 ms).

Due to the multi-input parameter switching behaviour of the device, we attempted to realize memtransistors that exhibit both electrical and optical plasticity. In its simplest form, such multi-input device could imitate the human visual synapse.[37] In the human visual system the retina plays a key role in light-sensing and visual perception through elaborate electrical and optical signal co-processing (**Figure 7.19a**). In our attempt to emulate the human retina, we approximated the "retina"-like synapse shown in **Figure 7.19b**. Here, rods and cones are denoted as a single type light-induced neurotransmitter. The bipolar and horizontal cells in the retina are labelled as "modulator", which introduces lateral inhibition to the system,[38] while **Figure 7.19c** shows the electric symbol of the bio-inspired memtransistor.

To test the operation of the bio-inspired memtransistor, input spikes were applied to the drain terminal and the excitatory postsynaptic current (EPSC) was measured at the source terminal (**Figure 7.19c**). To identify the appropriate potential voltage (V_{Pot}) and depression voltage (V_{Dep}), pinched hysteresis loops with different V_{DS} sweep range (from ±20 to ±80 V) were measured in the dark (**Figure 7.19d**). Upon application of a maximum-sweep voltage of ±20 V, the *p-n* doping configuration of the device is retained. When the V_{DS} sweep-range increases to ±40 V, the *p-n* junction configuration changes (see direction of hysteresis in **Figure 7.19d**). Thus, we set the V_{Pot} at the critical value of -20 V.

Figure 7.19e shows the evolution of EPSC as a function of the number of electrical pulses applied at the input at different V_G (0 to -20 V) measured in the dark. The postsynaptic current increases when applying a V_{Pot} and decreases when applying a V_{Dep}

(V_G= -20 V), demonstrating that synaptic plasticity, long-term potentiation (LTP) and long-term depression (LTD), are accelerated by applying a negative V_G, corresponding to the consolidation of long-term memory.[39,40] The behaviour is the result from the aforementioned gate field-induced charge carrier density increase. Illuminating the device with light (**Figure 7.19f**) results to EPSC that saturates faster when the V_G is set to -20 V than when it is set to 0 V, even upon the application of identical input spikes (V_{Pot} = -20 V, t_w=500 ms, t_i=200 ms). The effects resemble the conversion from short-term potentiation (STP) to long-term potentiation (LTP) by a combination of both light and electrical field, successfully mimicking the function of a neuromodulator. The ability to accelerate the modulation of the synaptic weight offer potential to improve the recognizing accuracy.[41] Our bio-inspired memtransistor may also be used to construct logic gates that are capable of processing different types of input signals, further complementing their analogue functions. Such capabilities can potentially be exploited to mimic the unconditioned reflex present in humans and animals.[42]

In conclusion, we demonstrated a novel artificial synapsis device architecture by integrating a memristor and a phototransistor using a hybrid heterostructure channel. The device is capable of co-process optical and electric signals mimicking certain functions of our visual system. The bio-inspired memtransistors exhibit heterosynaptic plasticity and multi-input modulation on plasticity weight. These capabilities can be potentially exploited for different applications including, neuromorphic vision systems, optoelectronic resistive random-access memory (OReRAM) and neuromorphic computing.[37]

7.2.6 Experimental Section

Materials preparation: Methylammonium Iodide (CH_3NH_3I, MAI, greatcell solar), Methylammonium Bromide (CH_3NH_3Br, MABr, greatcell solar) and PbI_2 (99.99%, trace metals basis, TCI), $PbBr_2$ (99.99%, trace metals basis, TCI) for 0.8M $MAPb(I_{1-x}Br_x)_3$ (x=0, 0.15, 1) precursor was stirred in a mixed organic solvent of anhydrous N,N-dimethylformamide (DMF, Sigma-Aldrich) and dimethyl sulfoxide (DMSO, Sigma-Aldrich) in a ratio of DMF : DMSO = 4 : 1 at 60 °C for 12 h. IDT-BT polymer was dissolved in chlorobenzene (10 mg/mL) in glovebox and stirred at 60 °C for 5 h. The synthesis details can be found elsewhere.

Device fabrication: the transistor is in a top-gate bottom-contact configuration. Au electrodes with a thickness of 40 nm were deposited on glass by thermal evaporator (Angstrom). The length (30 μm) and width (1 cm) of the device were defined by a shadow mask. Before the deposition of perovskite layer, the substrates were treated in UV-ozone. The perovskite precursor was spin-coated on the pre-treated substrates by a consecutive two-step spin-coating at 2,000 and 5,000 r.p.m for 10 and 20 s, respectively. 10 s before the end of the process, 400 μL of chlorobenzene was drop-cast on the spinning substrate (2 cm× 2 cm). The substrates were then annealed at 100 °C for 45 mins. The IDT-BT solution was spin-coated on the PVK layer at 2000 r.p.m for 60 s followed by an annealing process at 100 °C for 5 min in nitrogen. CYTOP was then spin-coated on the bilayer at

2000 r.p.m for 60 s. Finally, Al gate electrodes were deposited by thermal evaporation (Angstrom).

Characterization: Electrical characterization of transistors was carried out using a Keysight 2912A Precision Source/Measure Unit (SMU) and a probe station (EVERBEING) in a nitrogen-filled glove box. The optoelectronic characterisation was carried out by measuring the current-voltage characteristics of the fabricated devices in dark environment and under various illumination intensities using light-emitting diodes (LEDs, Thorlabs). Specifically, LEDs with electroluminescence emission peaks at 632 nm were employed. Various illumination intensities were achieved by a T-cube driver (Thorlabs). The actual illumination optical powers were calibrated by a Thorlabs 120UV power sensors prior to the measurement.

Molecular weight analysis: Number average (Mn) and weight-average (Mw) molecular weight were determined by Agilent Technologies 1200 series GPC running in chlorobenzene at 80 °C, using two PL mixed B columns in series, and calibrated against narrow polydispersity polystyrene standards. Photo *Electron Spectroscopy in Air (PESA)*: PESA spectra were recorded with a Riken Keiki AC-2 PESA spectrometer with a power setting of 5nW and a power number of 0.2. Samples for PESA were prepared on glass substrates by spin-coating. The ultraviolet-visible (UV-Vis) transmission measurements were performed using a Cary-5000 UV–Vis spectrophotometer (Agilent).

Optical spectroscopy measurements: Steady-state photoluminescence spectra were recorded using an excitation wavelength of 633 nm on a commercial spectrofluorometer

(Horiba). For TR-PL experiments samples were excited with the Coherent Helios 532 nm nanosecond laser with a pulse width of 0.85 nanoseconds and with a repetition rate of 1 kHz. Typical pulse energies were in the range of several μJ. The PL of the samples was collected by an optical telescope (consisting of two plano-convex lenses), and it was further focused on the slit of a spectrograph (PI Spectra Pro SP2300), and eventually detected with a Streak Camera (Hamamatsu C10910) system with a temporal resolution of 1.4 ps (picosecond). The data was acquired in photon counting mode using the Streak Camera software (HPDTA).

Atomic force microscope (AFM) analysis: AFM measurements were carried out with a Dimension Icon (Bruker) in air and in the dark. SKPFM measurement was carried out with a scanning probe microscope (Solver next, NT-MDT), and OSCM-PT probes (Veeco) were used in the SKPFM measurements. A Nova Nano- SEM 630 (FEI) was used to acquire cross-sectional images.

Grazing-incidence wide-angle X-ray scattering (GIWAXS): The GIWAXS experiments were conducted at CMS beamline (11BM) on NSLS II, Brookhaven National Lab. The wavelength of the X-ray was 1.17 Å with a bandwidth of 0.7%. The scattering signal was collected by a CCD detector, which was placed 220mm away from the sample with a tilt angle of 19° respect to the X-ray beam. The exposure time was 10s. The data analysis was performed by SciAnalysis.

References

1 Lee, Y. & Lee, T. W. Organic Synapses for Neuromorphic Electronics: From Brain-Inspired Computing to Sensorimotor Nervetronics. *Acc. Chem. Res.* **52**, 964-974 (2019).

2 Abbott, L. F. & Regehr, W. G. Synaptic computation. *Nature* **431**, 796-803 (2004).

3 Yang, J. J. S., Strukov, D. B. & Stewart, D. R. Memristive devices for computing. *Nat. Nanotechnol.* **8**, 13-24 (2013).

4 Kim, K. M. *et al.* Low-Power, Self-Rectifying, and Forming-Free Memristor with an Asymmetric Programing Voltage for a High-Density Crossbar Application. *Nano Lett.* **16**, 6724-6732 (2016).

5 Kuzum, D., Yu, S. M. & Wong, H. S. P. Synaptic electronics: materials, devices and applications. *Nanotechnology* **24**, 382001 (2013).

6 Sangwan, V. K. *et al.* Multi-terminal memtransistors from polycrystalline monolayer molybdenum disulfide. *Nature* **554**, 500-504 (2018).

7 John, R. A. *et al.* Ionotronic Halide Perovskite Drift-Diffusive Synapses for Low-Power Neuromorphic Computation. *Adv. Mater.* **30**, 1805454 (2018).

8 Prezioso, M. *et al.* Training and operation of an integrated neuromorphic network based on metal-oxide memristors. *Nature* **521**, 61-64 (2015).

9 Upadhyay, N. K. *et al.* Emerging Memory Devices for Neuromorphic Computing. *Advanced Materials Technologies* **4**, 1800589 (2019).

10 Fuller, E. J. *et al.* Parallel programming of an ionic floating-gate memory array for scalable neuromorphic computing. *Science* **364**, 570-574 (2019).

11 van de Burgt, Y. *et al.* A non-volatile organic electrochemical device as a low-voltage artificial synapse for neuromorphic computing. *Nat. Mater.* **16**, 414-418 (2017).

12 Choi, S. *et al.* SiGe epitaxial memory for neuromorphic computing with reproducible high performance based on engineered dislocations. *Nat. Mater.* **17**, 335-340 (2018).

13 Herz, L. M. Charge-Carrier Mobilities in Metal Halide Perovskites: Fundamental Mechanisms and Limits. *Acs Energy Letters* **2**, 1539-1548 (2017).

14 Jeon, N. J. *et al.* Solvent engineering for high-performance inorganic-organic hybrid perovskite solar cells. *Nat. Mater.* **13**, 897-903 (2014).

15 Green, M. A., Ho-Baillie, A. & Snaith, H. J. The emergence of perovskite solar cells. *Nat. Photonics* **8**, 506-514 (2014).

16 Ahmadi, M., Wu, T. & Hu, B. A Review on Organic-Inorganic Halide Perovskite Photodetectors: Device Engineering and Fundamental Physics. *Adv. Mater.* **29**, 1605242 (2017).

17 Wang, N. N. *et al.* Perovskite light-emitting diodes based on solution-processed self-organized multiple quantum wells. *Nat. Photonics* **10**, 699-704 (2016).

18 Yoo, E. J. *et al.* Resistive Switching Behavior in Organic-Inorganic Hybrid CH3NH3PbI3-xClx Perovskite for Resistive Random Access Memory Devices. *Adv. Mater.* **27**, 6170-6175 (2015).

19 Xiao, Z. G. *et al.* Giant switchable photovoltaic effect in organometal trihalide perovskite devices. *Nat. Mater.* **14**, 193-198 (2015).

20 Senanayak, S. P. *et al.* Understanding charge transport in lead iodide perovskite thin-film field-effect transistors. *Science Advances* **3**, e1601935 (2017).
21 Calado, P. *et al.* Evidence for ion migration in hybrid perovskite solar cells with minimal hysteresis. *Nat. Commun.* **7**, 13831 (2016).
22 Kim, G. Y. *et al.* Large tunable photoeffect on ion conduction in halide perovskites and implications for photodecomposition. *Nat. Mater.* **17**, 445-449 (2018).
23 Feldmann, J., Youngblood, N., Wright, C. D., Bhaskaran, H. & Pernice, W. H. P. All-optical spiking neurosynaptic networks with self-learning capabilities. *Nature* **569**, 208-214 (2019).
24 Johnston, M. B. & Herz, L. M. Hybrid Perovskites for Photovoltaics: Charge-Carrier Recombination, Diffusion, and Radiative Efficiencies. *Acc. Chem. Res.* **49**, 146-154 (2016).
25 D'Innocenzo, V. *et al.* Excitons versus free charges in organo-lead tri-halide perovskites. *Nat. Commun.* **5**, 3586 (2014).
26 Venkateshvaran, D. *et al.* Approaching disorder-free transport in high-mobility conjugated polymers. *Nature* **515**, 384-388 (2014).
27 Zhang, W. M. *et al.* Indacenodithiophene Semiconducting Polymers for High-Performance, Air-Stable Transistors. *J. Am. Chem. Soc.* **132**, 11437-11439 (2010).
28 Ma, C. *et al.* Heterostructured WS2/CH3NH3PbI3 Photoconductors with Suppressed Dark Current and Enhanced Photodetectivity. *Adv. Mater.* **28**, 3683-3689 (2016).
29 Lin, Y. H., Pattanasattayavong, P. & Anthopoulos, T. D. Metal-Halide Perovskite Transistors for Printed Electronics: Challenges and Opportunities. *Adv. Mater.* **29**, 1702838 (2017).
30 Zheng, G. H. J. *et al.* Manipulation of facet orientation in hybrid perovskite polycrystalline films by cation cascade. *Nat. Commun.* **9**, 2793 (2018).
31 Eames, C. *et al.* Ionic transport in hybrid lead iodide perovskite solar cells. *Nat. Commun.* **6**, 7497 (2015).
32 Li, C., Guerrero, A., Huettner, S. & Bisquert, J. Unravelling the role of vacancies in lead halide perovskite through electrical switching of photoluminescence. *Nat. Commun.* **9**, 5113 (2018).
33 Wei, L. Y., Ma, W., Lian, C. & Meng, S. Benign Interfacial Iodine Vacancies in Perovskite Solar Cells. *J. Phys. Chem. C* **121**, 5905-5913 (2017).
34 Zhu, X. J., Lee, J. & Lu, W. D. Iodine Vacancy Redistribution in Organic-Inorganic Halide Perovskite Films and Resistive Switching Effects. *Adv. Mater.* **29**, 1700527 (2017).
35 Tanaka, M. & Tachibana, M. Independent control of reciprocal and lateral inhibition at the axon terminal of retinal bipolar cells. *Journal of Physiology-London* **591**, 3833-3851 (2013).
36 Wang, Y. *et al.* Photonic Synapses Based on Inorganic Perovskite Quantum Dots for Neuromorphic Computing. *Adv. Mater.* **30**, 1802883 (2018).
37 Huh, W. *et al.* Synaptic Barristor Based on Phase-Engineered 2D Heterostructures. *Adv. Mater.* **30**, 1801447 (2018).
38 Zhou, F. C. *et al.* Optoelectronic resistive random access memory for neuromorphic vision sensors. *Nat. Nanotechnol.* **14**, 776-782 (2019).

Chapter *8*

Conclusion and Perspectives

This book was focused on our original research on the fabrication and characterization of perovskite-based optoelectronic devices, ranging from solar cells, photoconductors, phototransistors, inverters and multi-input parameter memtransistors. The main purpose is to interpret the underlying physics governing the carrier transport, and to optimize the device performance, through integrating with different dimensional materials (*i.e.* 0D, 1D, 2D, and 3D).

8.1 Conclusion

8.1.1 Photon Management with 0D Plasmonic Structure

In summary, the broad light absorption enhancement in the perovskite solar cells have been achieved by incorporating the plasmonic Au dimers core-shell nanoparticles into organometal halide perovskite solar cells. Due to the broadband optical absorption enhancement in incorporated perovskite, especially in the 550-800 nm spectral range, a 19.16% enhancement on conversion efficiency has been achieved at optimized addition ratio. Plasmonic structure is widely applied in photovoltaic, however the enhancement of conversion efficiency is limited (always below 15%), which results from the narrow resonance of the conventional nanostructures. Broadband enhancement is desired for the new generation of solar cells. Moreover, we systematically investigated the optical properties of the Au dimers by numerical simulation, which perfect matches our experimental results. With the superb properties of improving the light absorption in the long wavelength region, it can be expected potential for further increasing the PCE of solar cells.[1]

8.1.2 Enhancement of Photo-Detectivity with 2D vdW Heterostructure

High-performance photoconductor based on the hybrid perovskite and 2D materials is fabricated and characterized for the first time. Due to the superior properties of both perovskite and WS_2, photoconductor in this work shows the highest performance on detectivity among all the planar structure perovskite photodetector. Remarkably, the photodetectors exhibit more reliable high performance on ON/OFF ratios (~10^5) and high responsivity (~17 A/W) under 10^{-4} mW/cm^2 irradiation power. Thanks to the high mobility of WS_2 film and the charges transport separation by built-in barrier, the response time of hybrid WS_2/perovskite photoconductors are enhanced by four orders (from tens of seconds to milliseconds). Furthermore, WS_2 film contributes as an atomically flat and highly ordered substrate, perovskite film grows on it shows better orientation and better crystalline properties, which are benefit to the electrons transport and result in a better charge transfer. WS_2 film can also be a superior passivation layer, which is benefit to the charge release from the interface, and then improves the response properties of photodetectors. At last, WS_2/perovskite induced built-in barrier provides a depletion region to inhabit the recombination of electron and holes, which lead to a better photoresponse and faster response speed. Our results for the first time built the high-performance TMD/perovskite heterojunction photoconductors, and provide a useful insight into charge transport in hybrid WS_2/perovskite system, promoting their future applications in optoelectronic devices.[2]

8.1.3 Enhancement of Carrier Transport with 1D Bulk Heterostructure

Semiconducting-enriched single-walled carbon nanotubes (95% s-SWCNT) have been used to enhance the performance of mixed-cation perovskite transistors and inverters, enabling solution-processed mixed-dimensional electronics with high mobility and low operation bias. The mixed-dimensional PVK/SWCNT heterostructures exhibited a low OFF current (10^{-11} A), leading to an ON/OFF ratio up to 10^7, which is the highest reported to date for solution-processed devices. Furthermore, we fabricated high-gain inverters with a switching voltage of 1 V, which provides opportunities for digital electronic applications. Finally, our temperature-dependent transport study revealed a transition from high-temperature ambipolar to low-temperature unipolar behavior, which is a result of competing carrier trapping and emission. The demonstrated merits of low-cost solution processing, high carrier mobility, low-voltage operation, high ON/OFF ratio, and versatile operation modes make mixed-dimensional PVK/SWCNT heterostructures a highly promising candidate for a variety of electronic applications.

8.1.4 Multi-Input Memtransistor with 3D Heterostructure

We demonstrated a new artificial synaptic architecture by integrating a memristor and a phototransistor using IDT-BT/PVK heterostructure, mimicking the visual system of human beings. The memtransistor could realize heterosynaptic plasticity and multi-input modulation on plasticity weight. This unprecedented device could be applied to optoelectronic resistive random-access memory (OReRAM) and neuromorphic computing. Furthermore, our demonstration reveals the ionic and electronic dynamics behind the devices.

8.2 Challenges and Perspectives

8.2.1 Toxicity

Lead (Pb), which consists a big portion of perovskites, is a toxic heavy metal. Addressing toxicity concerns is essential in industry to ensure the widespread use and possible incorporated with other devices. However, such studies are lacking due to the outstanding performance of Pb perovskite devices. Some researchers tried to replace the Pb with Bismuth (Bi)[3] and Tin (Sn).[4] Although the device performance is not able to compete with that with Pb, a mixture of the elements could open a door for the non-(less-) toxic devices.

8.2.2 Stability

It has been studied that MA+ group and Iodine atoms in perovskites give rise to the ion migration, hysteresis.[5] Besides, moisture in ambient condition also harmful for the operation of perovskite devices. Approaches to encapsulate the perovskite employing hydrophobic molecules or directly incorporating a hydrophobic element in the device have provided progress towards maintaining high performances.[6]

8.2.3 2D Perovskites

Low dimensional halide perovskites have been attracting intensive attention in recent years, including perovskite nanowires,[7] 2D perovskite nanoplates.[8] Various methods were studied to fabricate light emitting diodes (LEDs),[9] and photodetectors.[10] When the dimension reduces, some novel properties would appear, such as tunable bandgap,[11] high photoluminescence quantum yield,[9] *etc.* In this regard, synthesis low dimensional perovskites and utilizing them in optoelectronics are anticipated in the future.

Reference

1 Ma, C. *et al.* Plasmonic - Enhanced Light Harvesting and Perovskite Solar Cell Performance Using Au Biometric Dimers with Broadband Structural Darkness. *Sol Rrl* **3**, 1900138 (2019).

2 Ma, C. *et al.* Heterostructured WS2/CH3NH3PbI3 Photoconductors with Suppressed Dark Current and Enhanced Photodetectivity. *Adv. Mater.* **28**, 3683-3689 (2016).

3 Yang, B. *et al.* Lead-Free, Air-Stable All-Inorganic Cesium Bismuth Halide Perovskite Nanocrystals. *Angew Chem Int Edit* **56**, 12471-12475 (2017).

4 Mitzi, D. B. Solution-processed inorganic semiconductors. *J. Mater. Chem.* **14**, 2355-2365 (2004).

5 Turren-Cruz, S. H., Hagfeldt, A. & Saliba, M. Methylammonium-free, high-performance, and stable perovskite solar cells on a planar architecture. *Science* **362**, 449 (2018).

6 Han, T. H. *et al.* Perovskite-polymer composite cross-linker approach for highly-stable and efficient perovskite solar cells. *Nat. Commun.* **10**, 520 (2019).

7 Deng, W. *et al.* Ultrahigh-Responsivity Photodetectors from Perovskite Nanowire Arrays for Sequentially Tunable Spectral Measurement. *Nano Lett.* **17**, 2482-2489 (2017).

8 Deng, W. *et al.* 2D Ruddlesden-Popper Perovskite Nanoplate Based Deep-Blue Light-Emitting Diodes for Light Communication. *Adv. Funct. Mater.* **29**, 1903861 (2019).

9 Zhao, B. D. *et al.* High-efficiency perovskite-polymer bulk heterostructure light-emitting diodes. *Nat. Photonics* **12**, 783 (2018).

10 Lee, H. J. *et al.* Ultrahigh-Mobility and Solution-Processed Inorganic P-Channel Thin-Film Transistors Based on a Transition-Metal Halide Semiconductor. *ACS Appl. Mater. Interfaces* **11**, 40243-40251 (2019).

11 Chen, L. J. *et al.* Wavelength-Tunable and Highly Stable Perovskite-Quantum-Dot-Doped Lasers with Liquid Crystal Lasing Cavities. *ACS Appl. Mater. Interfaces* **10**, 33307-33315 (2018).

Appendices

[1] **Ma, C**., Chen, H., Yengel, E., Faber, H., Khan, J., Zhang, W., Laquai, F., McCulloch, I. & Anthopoulos, T. D. Multi-Input Parameter Modulable Memtransistors from Hybrid Perovskite/Conjugated Polymer Heterostructures for neuromorphic computing. Revision in *Nature Electronics*.

[2] **Ma, C**., Clark, S., Liu, Z., Liang, L., Tao, R., Han, A., Liu, X., Li, L.-J., Anthopoulos, T. D., Hersam, M. C. & Wu, T. Solution-Processed Mixed-Dimensional Hybrid Perovskite/Carbon Nanotube Optoelectronics.
ACS Nano, DOI: 10.1021/10.1021/acsnano.9b07888.

[3] **Ma, C.,** Liu, C., Huang, J., Ma, Y., Liu, Z., Li, L.-J., Anthopoulos, T. D., Han, Y., Fratalocchi, A., & Wu, T. Plasmonic-enhanced light harvesting and perovskite solar cell performance using Au biometric dimers with broadband structural darkness. ***Solar RRL***, 3(8), 1900138.

[4] **Ma, C**., Shi, Y., Hu, W., Chiu, M. H., Liu, Z., Bera, A., ... & Wu, T. (2016). Heterostructured WS2/CH3NH3PbI3 photoconductors with suppressed dark current and enhanced photodetectivity. *Advanced Materials*, 28(19), 3683-3689.

[5] Li, F., **Ma, C.**, Wang, H., Hu, W., Yu, W., Sheikh, A. D., & Wu, T. (2015). Ambipolar solution-processed hybrid perovskite phototransistors. *Nature communications*, 6, 8238.

[6] Wang, H., Min, S., **Ma, C.,** Liu, Z., Zhang, W., ... & Zhu, H. (2017). Synthesis of single-crystal-like nanoporous carbon membranes and their application in overall water splitting. *Nature communications*, 8, 13592.

[7] Yu, W., Li, F., Yu, L., Niazi, M. R., Zou, Y., Corzo, D., **Ma. C**., ... & Amassian. A. (2018). Single crystal hybrid perovskite field-effect transistors. *Nature communications*, 9(1), 5354.

[8] Lin, Y.-H., Huang, W., … **Ma, C**., …, & Snaith, H, J. (2019). Deciphering photocarrier dynamics for tuneable high-performance perovskite-organic semiconductor heterojunction phototransistors. *Nature communications*, 10, 4475.

[9] Li, Z., Li, H., …**Ma, C**., …, Anthopoulos, T. D. & Shi, Y. (2019) Self-powered perovskite/CdS heterostructure photodetectors. ***ACS applied materials interfaces***.

[10] Han, A., Aljarb, A., … **Ma, C**., … Anthopoulos, T. D. & Li, L.-J. (2019) Growth of 2H stacked WSe_2 bilayers on sapphire. *Nanoscale horizon*, **4**, 1434.

[11] Chen, H., Wadsworth, A., **Ma, C**., … Sirringhaus, H. & McCulloch, I. (2019) The effect of ring expansion in Thienobenzo[b]indacenodithiophene polymer for organic field effect transistors. *Journal of the American Chemical Society*.

[12] Peng, W., Wang, L., Murali, B., Ho, K. T., Bera, A., Cho, N., ... **Ma, C.,** ... &Bakr, O. (2016). Solution-Grown Monocrystalline Hybrid Perovskite Films for Hole-Transporter-Free Solar Cells. *Advanced Materials*, 28(17), 3383-3390.

[13] Li, F., Wang, H., Kufer, D., Liang, L., Yu, W., Alarousu, E., **Ma. C**... & Wu. T (2017). Ultrahigh Carrier Mobility Achieved in Photoresponsive Hybrid Perovskite Films via Coupling with Single-Walled Carbon Nanotubes. *Advanced Materials*, 29(16), 1602432.

[14] Wang, H., Sheikh, A. D., Feng, Q., Li, F., Chen, Y., Yu, W., **Ma. C**... & Wu. T (2015). Facile synthesis and high performance of a new carbazole-based hole-transporting material for hybrid perovskite solar cells. *ACS Photonics*, 2(7), 849-855.

[15] Wang, H., Min, S., Wang, Q., Li, D., Casillas, G., **Ma, C**., ... & Wu. T (2017). Nitrogen-doped nanoporous carbon membranes with Co/CoP Janus-type nanocrystals as hydrogen evolution electrode in both acidic and alkaline environments. *ACS nano*, 11(4), 4358-4364.

[16] Wang, L., Ma, H., Chang, L., **Ma, C**., Yuan, G., Wang, J., & Wu, T. (2017). Ferroelectric BiFeO3 as an Oxide Dye in Highly Tunable Mesoporous All-Oxide Photovoltaic Heterojunctions. *Small*, 13(1), 1602355.

[17] Wang, L., Li, Y., Bera, A., **Ma, C.,** Jin, F., Yuan, K., ... & Wu. T. (2015). Device performance of the Mott insulator LaVO 3 as a photovoltaic material. *Physical Review Applied*, 3(6), 064015.

[18] Wang, H., Jia, J., Song, P., Wang, Q., Li, D., Min, S., ... **Ma. C** & Wu, T. (2017). Efficient Electrocatalytic Reduction of CO2 by Nitrogen-Doped Nanoporous Carbon/Carbon Nanotube Membranes: A Step Towards the Electrochemical CO2 Refinery. *Angewandte Chemie*, 129(27), 7955-7960.

[19] Wu, K., Bera, A., **Ma, C.,** Du, Y., Yang, Y., Li, L., & Wu, T. (2014). Temperature-dependent excitonic photoluminescence of hybrid organometal halide perovskite films. *Physical Chemistry Chemical Physics*, 16(41), 22476-22481.

[20] Li, F., Chen, Y., **Ma, C**., Buttner, U., Leo, K., & Wu, T. (2017). High-Performance Near-Infrared Phototransistor Based on n-Type Small-Molecular Organic Semiconductor. *Advanced Electronic Materials*, 3(1), 1600430.

[21] Liu, Z., Li, Y., Guan, X., Mi, Y., Al-Hussain, A., Ha, S. T., Chiu, M. **Ma, C**. ... & Wu, T. (2019). One-Step Vapor-Phase Synthesis and Quantum-Confined Exciton in Single-Crystal Platelets of Hybrid Halide Perovskites. *The Journal of Physical Chemistry Letters*, 10, 2363-2371.

www.ingramcontent.com/pod-product-compliance
Lightning Source LLC
LaVergne TN
LVHW091422190726
843491LV00006B/1553

* 9 7 8 3 3 8 4 2 5 3 7 4 3 *